M000236041

Bushmeat and Livelihoods

Conservation Science and Practice Series
Published in association with the Zoological Society of London

Blackwell Publishing and the Zoological Society of London are proud to present our new *Conservation Science and Practice* volume series. Each book in the series reviews a key issue in conservation today. We are particularly keen to publish books that address the multidisciplinary aspects of conservation, looking at how biological scientists and ecologists are interacting with social scientists to effect long-term, sustainable conservation measures.

Books in the series can be single or multi-authored and proposals should be sent to:

Ward Cooper, Senior Commissioning Editor, Blackwell Publishing Ltd, 9600 Garsington Road, Oxford OX4 2DQ, UK
Email: ward.cooper@oxon.blackwellpublishing.com

Each book proposal will be assessed by independent academic referees, as well as our Series Editorial Panel. Members of the Panel include:

Richard Cowling, Nelson Mandela Metropolitan, Port Elizabeth, South Africa
John Gittleman, Institute of Ecology, University of Georgia, USA
Andrew Knight, Nelson Mandela Metropolitan, Port Elizabeth, South Africa
Georgina Mace, Imperial College London, Silwood Park, UK
Daniel Pauly, University of British Columbia, Canada
Stuart Pimm, Duke University, USA
Hugh Possingham, University of Queensland, Australia
Peter Raven, Missouri Botanical Gardens, USA
Michael Samways, University of Stellenbosch, South Africa
Nigel Stork, University of Melbourne, Australia
Rosie Woodroffe, University of California, Davis, USA

Conservation Science and Practice Series

Bushmeat and Livelihoods: Wildlife Management and Poverty Reduction

Edited by

Glyn Davies

David Brown

© 2007 by Blackwell Publishing Ltd

BLACKWELL PUBLISHING

350 Main Street, Malden, MA 02148-5020, USA
9600 Garsington Road, Oxford OX4 2DQ, UK
550 Swanston Street, Carlton, Victoria 3053, Australia

The right of Glyn Davies and David Brown to be identified as the Authors of the Editorial Material in this Work has been asserted in accordance with the UK Copyright, Designs, and Patents Act 1988.

All rights reserved. No part of this publication may be reproduced, stored in a retrieval system, or transmitted, in any form or by any means, electronic, mechanical, photocopying, recording or otherwise, except as permitted by the UK Copyright, Designs, and Patents Act 1988, without the prior permission of the publisher.

Designations used by companies to distinguish their products are often claimed as trademarks. All brand names and product names used in this book are trade names, service marks, trademarks, or registered trademarks of their respective owners. The publisher is not associated with any product or vendor mentioned in this book.

This publication is designed to provide accurate and authoritative information in regard to the subject matter covered. It is sold on the understanding that the publisher is not engaged in rendering professional services. If professional advice or other expert assistance is required, the services of a competent professional should be sought.

First published 2007 by Blackwell Publishing Ltd

1   2007

Library of Congress Cataloging-in-Publication Data

Bushmeat and livelihoods : wildlife management and poverty reduction / edited by Glyn Davies and David Brown.
     p. cm. – (Conservation science and practice series : v. 2)
   "Published in association with the Zoological Society of London."
   Includes bibliographical references and index.
   ISBN 978-1-4051-6779-6 (pbk. : alk. paper) 1. Wildlife as food~Africa. 2. Wildlife as food–Economic aspects–Africa. 3. Wildlife management–Africa. I. Davies, Glyn. II. Brown, David, 1945– III. Zoological Society of London.

   SK571.B87 2007
   333.95′413096–dc22

                   2007016683

A catalogue record for this title is available from the British Library.

Set in 10.5/12.5 Minion
by Prepress Projects Ltd, Perth, Scotland
Printed and bound in Singapore
by Fabulous Printers Pte Ltd

The publisher's policy is to use permanent paper from mills that operate a sustainable forestry policy, and which has been manufactured from pulp processed using acid-free and elementary chlorine-free practices. Furthermore, the publisher ensures that the text paper and cover board used have met acceptable environmental accreditation standards.

For further information on
Blackwell Publishing, visit our website:
www.blackwellpublishing.com

Contents

Contributors

Eric Arancibia
Calle Lilia Salvatierra 4660
Urbanizacion Banunion
Santa Cruz de la Sierra
Bolivia

Elizabeth L. Bennett
Wildlife Conservation Society
2300 Southern Blvd
Bronx
New York 10460
USA

Neil Bird
Overseas Development Institute
111 Westminster Bridge Road
London SE1 7JD
UK

David Brown
Overseas Development Institute
111 Westminster Bridge Road
London SE1 7JD
UK

Taylor Brown
theIDLgroup
Brockley Combe
Backwell
Bristol BS48 3DF
UK

Connie J. Clark
School of Natural Resources and
Environment and Department of
Zoology
PO 11852

University of Florida
Gainesville
FL 32611-8525
USA
and
Wildlife Conservation Society
BP 14537
Brazzaville
Republic of Congo

Guy Cowlishaw
Institute of Zoology
Zoological Society of London
Regents Park
London NW1 4RY
UK

Glyn Davies
Conservation Programmes
Zoological Society of London
Regent's Park
London NW1 4RY
UK

Chris Dickson
Overseas Development Institute
111 Westminster Bridge Road
London SE1 7JD
UK

Tamsyn East
Division of Biology
Imperial College London
Silwood Park Campus
Ascot SL5 7PY
UK

Fabrice Edouard
Apartado Postal No. 2
Centro Comercial Reforma
CP 68051
Oaxaca
Estado de Oaxaca
Mexico

John E. Fa
Durrell Wildlife Conservation Trust
Les Augres Manor
Trinity
Jersey JE3 5BP
UK

Andrew Hurst
Overseas Development Institute
111 Westminster Bridge Road
London SE1 7JD
UK

Catarina Illsley
Grupo de Estudios
Ambientales, A C
Allende 7
Santa Ursula Coapa
Mexico

Nick Keylock
Division of Biology
Imperial College London
Silwood Park Campus
Ascot SL5 7PY
UK

Noëlle F. Kümpel
Division of Biology
Imperial College London
Manor House
Silwood Park Campus
Ascot SL5 7PY
UK
and
Conservation Programmes
Zoological Society of London
Regent's Park
London NW1 4RY
UK

Andrew Long
Department for International
Development (DFID)
1 Palace Street
London SW1E 5HE
UK

Stuart A. Marks
Mipashi Associates
10 Ashwood Square
Durham
NC 27713
USA

Elaine Marshall
UNEP-WCMC
219c Huntingdon Road
Cambridge CB3 0DL
UK

Germain A. Mavah
Wildlife Conservation Society
BP 14537
Brazzaville
Republic of Congo

Samantha Mendelson
Conservation Programmes/
Institute of Zoology
Zoological Society of London
Regent's Park
London NW1 4RY
UK

E. J. Milner-Gulland
Division of Biology
Imperial College London
Silwood Park Campus
Ascot SL5 7PY
UK

Adrian Newton
School of Conservation Sciences
Bournemouth University
Talbot Campus
Poole BH12 5BB
UK

John R. Poulsen
Department of Zoology
PO 11852, University of Florida
Gainesville
FL 32611–8525
USA
and
Wildlife Conservation Society
BP 14537
Brazzaville
Republic of Congo

J. Marcus Rowcliffe
Institute of Zoology
Zoological Society of London
Regents Park
London NW1 4RY
UK

Jonathan Rushton
CEVEP
Casilla 10474
La Paz
Bolivia

Kathrin Schreckenberg
Overseas Development Institute
111 Westminster Bridge Road
London SE1 7JD
UK

Björn Schulte-Herbrüggen
Conservation Programmes
Zoological Society of London
Regent's Park
London NW1 4RY
UK

Hilary Solly
GEPAC (Gestion Participative en
Afrique Centrale)
Centre d'Anthropologie Sociale et
Culturelle
and
Université Libre de Bruxelles
44 Avenue Jeanne
1050 Brussels
Belgium

Ross C. Thompson
Beverly and Qamanirjuaq Caribou
Management Board
Box 629
Stonewall
Manitoba R0C 2Z0
Canada

Christopher Vaughan
Whiteknights
PO Box 217
Reading RG6 6AH
UK

Dirk Willem te Velde
Overseas Development Institute
111 Westminster Bridge Road
London SE1 7JD
UK

Preface

This volume arose out of an international conference on Bushmeat and Livelihoods held at the Zoological Society of London (ZSL) on 23–24 September 2004. The aim was to address the bushmeat issue in ways that would be of interest to conservationists and development practitioners. The conference was organized by the ZSL and the Overseas Development Institute (ODI), with funding from ZSL's Bushmeat and Forests Conservation Programme and ODI's Wild Meat, Livelihoods Security and Conservation in the Tropics project. The latter was supported by the John D. and Catherine T. MacArthur Foundation, which is gratefully acknowledged for its financial support for the project and its contribution towards the funding of the ZSL conference.

The conference took stock of the growing interest in the bushmeat trade as a potential threat to wildlife conservation in the tropics, yet with the simultaneous recognition that bushmeat is also a significant component of the livelihood strategies for peoples of the producer states. Thus, the 'bushmeat crisis' is not amenable to resolution by conservation strategies alone. The conference sought to bring together both streams of research so as to recommend ways forward for both policy-makers and practitioners. This required adopting both biological and socioeconomic perspectives and reflecting on policy frameworks and entry points for policy development.

A wide range of professionals, upwards of 70 researchers, policy-makers and practitioners, working on bushmeat and related issues came together at the conference, which provided an exceptional opportunity to explore the differences in perspective that arise when the human dimension is brought to the fore and priority is given to livelihoods, economic and sociocultural concerns. Coverage was pan-tropical, with much concentration on West Africa and the Congo Basin, in line with the dominance of the subregion in the bushmeat trade. The aim was to share experience both intraregionally and across continents (Africa, Latin America-Caribbean, Asia-Pacific), and to identify the lessons that might be learned from relevant extrasectoral experiences. The main, but not the only, emphasis was on mammals in tropical moist forests, for it is here that the conservation challenges are greatest.

All but one of the chapters in this volume were first presented at the conference, and subsequently revised for inclusion in this work. We much appreciate the contribution of the authors, and we would also like to acknowledge the contribution of our two subeditors, Camilla Fritze and Chris Ransom, and the support we received from Zoe Parr and Jane Loveless in organizing the original conference.

Glyn Davies and David Brown
London

Introduction

David Brown and Glyn Davies

'Bushmeat' is a topical issue in conservation circles. Generally taken to mean the meat of wild mammals (though sometimes seen as including all wild terrestrial mammals, birds, reptiles and insects), it has also been referred to as 'game meat' in savanna regions of Africa. Bushmeat is traded raw, or in a smoked and preserved form, and is a key dimension of the livelihoods of forest dwellers, and an important item of consumption for the populations of many forest-rich countries. It is also under threat in conservation terms. If off-take continues at the present high levels in key areas, then some important and charismatic biodiversity will be lost.

Bushmeat is unusual among human foodstuffs in that it excites interest and high emotions in countries far removed from the sites of its consumption. This is unusual in two respects: first, the fact that it is local food items which are of interest and, second, the fact that it is non-consumers who, by and large, are the most vocal parties. The other tropical products that excite high emotions are often ones in which northern consumers have a direct interest. Illegal timber is a good example. In the case of bushmeat, the primary interests appear to be in the conservation of wildlife populations, the welfare of the species consumed and their existence values, as indicators of healthy ecosystems which appear threatened as never before. It is seldom the interests of the humans who consume them.

The scale of the bushmeat trade

Before entering the debate about bushmeat and livelihoods, the scale of the issue needs to be understood. The levels of off-take vary by ecological zone, country and continent, but by and large levels of off-take are highest in the humid forests of West–Central Africa, and lower (though still significant) in Asia and South America. One of the earliest assessments of the scale of bushmeat/game consumption (including sport hunting and ranching) was made in the early 1990s. It noted that the annual consumption of 'game meat' in Africa was over a million tonnes, and stressed that wildlife had become a 'forgotten resource' (Chardonnet *et al.*, 2002).

More recent estimates of bushmeat consumption have been at least as high, and sometimes considerably higher. Figures for the Congo Basin, for example, range between one and five million tonnes per annum (Wilkie and Carpenter,

1999; Fa and Peres, 2001; Fa et al., 2003). Such levels are quite compatible with the estimates for the economic values of traded goods made in a number of national studies in the West–Central African subregion.

The value of national bushmeat trade in some West and Central African countries (after Davies, 2002).

Country	Estimated value of trade (US$m/annum)
Gabon	26 urban; 22 rural
Central African Republic	22 (production)
Cote d'Ivoire	105
Ghana	205
Liberia	42

A number of factors account for the differences in the levels of consumption and trade between tropical continents. Africa's moist forests are generally much more productive than those of Asia and South America, particularly as regards terrestrial mammals (Fa and Peres, 2001). An additional reason for the high dependence on bushmeat in West and Central Africa is that supplies of domestic animal protein are generally very limited, due to the prevalence of disease and the high costs of husbandry (Part 4). This is particularly so in areas with low human populations, where labour tends to be unspecialized and infrastructure (for example, fencing to keep grazers off farmland) is lacking. This is compounded by the fact that capital investment in African agribusiness has generally been much lower than in the other tropics, most notably in Brazil, which also has diminishing supplies of domestic animal protein.

Africa's dryland forests are more productive than its moist forests, particularly as regards ungulates, but for precisely this reason they tend also to be more productive of domestic livestock, and hence proportionately less involved in the provision of wild animal protein.

It can thus be established that bushmeat is a valued food item in the tropical producer countries, particularly in West–Central Africa. However, additional steps are needed to establish the extent to which it is also a key dimension of the livelihoods of forest-dependent peoples, rather than simply part of a package of natural resources which can be made use of, as circumstances permit. Answering this question – even attempting to interpret the evidence which is already available – takes us into a more contentious area, at the boundary between conservation and development.

Bushmeat in conservation and development

Until recently, most research on the bushmeat issue has been driven by conservation priorities, and livelihoods concerns have tended to be secondary and contingent. However, the importance of bushmeat in range state economies requires that policy development also takes the human dimension fully into account. Equally, if bushmeat projects are to link up with the main themes of international development policy, then a more socioeconomic focus is needed. This involves situating the hunting and the harvest of wild meat within the setting of the rural economy, and understanding the linkages between local livelihoods and national and regional patterns of trade. The starting point must be the actual dynamics of the bushmeat economy, as a productive sector, and the role which a more managed and controlled off-take might have in both conserving the resource and providing for the welfare of the human populations that depend on it.

For these two aims to be achieved, the livelihoods dimensions must be set within the framework of development strategy. This implies more information on such issues as:

- the nature and extent of the poverty linkages in bushmeat production and consumption, including the distribution of value in the bushmeat commodity chain;
- the policy constraints that influence the trade, and their implications for the welfare of the poor;
- the underlying governance issues, and the potential for these to be addressed within the wildlife subsector;
- the contribution that the wildlife sector can make to socioeconomic development, and the trade-offs between development and conservation;
- the lessons that might be learnt from other disciplines as regards the reconciliation of social and conservation aims in policy development.

This is a particularly timely moment to address such issues as there is a significant volume of important new research work (much of it represented in this volume) which is seeking to better understand the social dimensions of the bushmeat economy, and thereby to reconcile conservation and development priorities. Indeed, the very fact that such a corpus of new work is developing is itself of interest. One of the most notable aspects of the recent debate on bushmeat is the extent of the convergence in discourse around the livelihoods and human welfare theme, and the shared needs of both development and conservation sectors for transparent and accountable governance (Davies, 2002).

At least as viewed from the northern research community, individuals and organizations of all persuasions now appear to be addressing the same set of issues, with the same desire to accommodate present-day realities and cultural relativities (e.g. Bennett *et al.*, 2007). All this is in marked contrast to the polarization and conflict which are characteristic of the treatment of the issue in the media, and the doomsday thinking which has hitherto beset the public presentation of the bushmeat issue, like many environmental issues in the developing world.

Policy framework

The pace of policy development specific to the bushmeat theme has also quickened in recent years. Milestones have included the creation of the CITES Sub-regional Working Group on Bushmeat in 2000, which followed endorsement of the UK submission to the Eleventh Conference of the Parties, 'Bushmeat as a Trade and Wildlife Management Issue' (doc. 11/44). This reported, on an interim basis, at CoP 12 in Santiago, and gave its final report in October 2004 at CoP 13 in Bangkok. The Convention on Biological Diversity has also pledged its support for 'the development of policies, enabling legislation and strategies that promote sustainable use of, and trade in, non-timber forest products, particularly bushmeat and related products' (CoP 6, decision VI/22). A number of papers on this theme are to be presented to the Convention's Subsidiary Body on Scientific, Technical and Technological Advice (SBSTTA) in the coming months.

The parallel European-wide policy process of supporting improved governance to reduce the illegal timber trade (Forest Law Enforcement and Governance against Illegal Timber; EC, 2005), along with the associated African Forest Law Enforcement and Governance Ministerial Declaration (Yaounde, 2003), has also drawn attention to the need to improve forest management as a whole, and in particular to focus on reducing wildlife killing as a by-product of timber production. This has led to an interest in introducing a wildlife component into timber certification processes, and using these to strengthen respect for wildlife laws and wildlife-friendly management practices (Schulte-Herbrüggen and Davies, 2006).

As regards the international dimension, there are therefore some promising developments. However, the biggest challenge still remains at the national policy level within the producer states. Attempts to advance international and regional policy from a range state perspective have had only limited success, and this can be put down largely to lack of direction at the national level. The CITES Central African Bushmeat Working Group has struggled to find common ground within the labyrinthine and outdated legislation of its member states, in ways which recognize the realities of the trade but which do not surrender conservation principles or give unwelcome signals to donors and paymasters

What is urgently required is an attempt to ground international concerns about the threat to biodiversity posed by the burgeoning bushmeat trade within the priorities of the producer states. This is likely to require innovative approaches to conservation practice, which recognize the important use values, as well as existence values, that tropical wildlife represents, and much greater consideration than hitherto of the means to bring these issues into national policy. This volume is an attempt to address both of these dimensions.

The evidence

If conservation is to be successfully reconciled with development, then the first step is the reconciliation of species-specific data with an understanding of the demand side, in terms of both livelihoods and markets. This is the subject of the first part of the book (Part 1: Bushmeat – Markets and Households). Important topics of debate in this subject area include the differences in species and hunting dynamics between farmbush and various forest habitats; the distribution of benefits within the trade; and the implications of both for conservation.

As Davies and Robinson note in their introduction to Part 1, our understanding of the volumes of bushmeat traded is still limited. However, whenever the topic has been studied in-depth, the extent of the dependence is almost always greater than had been assumed, often notably so. How to interpret high dependence on bushmeat is an additional area of complexity. Participants at the conference were divided on the significance. For example, what are the implications of a 10% average dependence? Is the implication that bushmeat is not significant, and could be easily replaced by other commodities, or that this is a vital protein source during periods of hardship? Aggregate dependence is often a poor indicator of cultural importance, as the UK and USA turkey industries attest. Cultural tastes and preferences may well be more tenacious than conservation planners would like to think, and cannot easily be gainsaid. As Davies remarks, it may be possible to capitalize on cultural preferences and taboos (as with the taboo against eating great apes in many parts of West Africa), but the reverse is not necessarily the case (Chapter 1). Breaking cultural preferences may be a much harder task, particularly where the incentives are just not there.

The answer to the high dependence conundrum is likely to be situationally determined. Where the poor are heavily dependent on a broad mix of livelihoods activities, none of them particularly productive but critical in their combination because of their risk mitigation effects, then 10% dependency is probably high, and it may be non-substitutable. Even where other alternatives seem feasible, economic realities may favour the continuance of the bushmeat trade.

For example, developing new fisheries has often been suggested as an alternative source of protein, but this may well be overoptimistic, particularly as regards freshwater fisheries. A sure sign of the potential of fisheries will be the presence of fishermen who are already there. Pristine fishing sites may well be pristine because of their lack of economic potential, not because their potential is untapped. In such situations, the heightened role that bushmeat plays is likely to be the product of sound economics, not lack of awareness of alternatives.

When we turn to the species composition of the trade, we are again confronted with considerable uncertainty. It is apparent that the profile of species that dominate the market trade is often quite at odds with public perceptions. Primates, especially the great apes, are less strongly represented in the trade than media headlines would suggest, and indeed are preferred in only a few areas. This does not imply that they are unthreatened by the trade, however. Opinions vary on whether discrimination in the hunt is a practicable ambition to regulate the trade. Opponents are sceptical; they argue that once hunters get a saleable animal in their sights, they will pull the trigger regardless. Proponents would argue that the lack of discrimination is but a symptom of the low governance of the subsector, and the lack of any incentives to discriminate. Where it is in the interests of hunters to refrain from the kill, then they will do so, and vice versa.

Where extinction thresholds for larger species have been passed, the situation is much less critical. The main message in such instances would appear to be where not to place conservation efforts, as much as where to concentrate them. The research undertaken in Takoradi market, by Cowlishaw *et al.* (Chapter 2), is particularly instructive here. Though a focus of campaign efforts in Ghana, the trade is almost exclusively orientated to fast-breeding 'pest' species, and the evidence is that these are neither endangered nor easy to control. As this trade is largely in the hands of small local hunters, not criminal gangs, it seems a waste of effort, and a dispersal of cultural capital, to campaign heavily against it. The situation in the national parks may be different, and merit much greater attention due to the threat to mammalian biodiversity. The future value of the tourist trade is an additional consideration here, and a topic which is more easily brought into national policy than more abstract concerns with biodiversity.

Other chapters in the first part of this volume reinforce the notion that a very varied range of species is found in different markets – primates and antelopes and rodents each occurring in different proportions – such as that by John Fa (Chapter 3). Furthermore, there is important evidence from Noëlle Kümpel *et al.* that fresh meat (whether domestic or bushmeat) is preferred over frozen items in an urban market in Equatorial Guinea (Chapter 5). The impact of market price in determining food choices is a recurrent theme in this part, but supply estimates could not be reliably determined from bushmeat market data, given that many animals are left in traps, consumed in forest camps or field huts, used in households and privately traded or gifted, and so never reach the market (Chapter 3). Two other chapters in this section take a more socioeconomic perspective of the trade in forest and savanna habitats, emphasizing the importance for income and food for various poor groups. Hilary Solly notes that stopping village hunters, who are not killing rare animals, in the nearby Dja Reserve in Cameroon only served to alienate them from the conservation effort because they had few alternatives for food or crop protection (Chapter 4). This theme was revisited in relation to poor farmers living in the plateau adjacent to the Luangwa Valley, Zambia. Here, Marks and Short record that cash needs are driving a marked increase in hunting (Chapter 6). Interestingly, Lewis (Chapter 11) provides evidence that supporting improved agricultural trading in the same region, albeit with substantial external start-up funds, has led to increases in wildlife numbers.

The second section of this volume (Part 2: Institutional Contexts) introduces the institutional dimension to the bushmeat and livelihoods debate. This is the most important interface between the national and the international agendas, and an area of tension in those societies where there has been a drive to internationalize the responsibilities. Opinions are divided as to whether bushmeat policy can be advanced more effectively through legislation or enforcement. Some would see legal reform as essential, though not necessarily sufficient on its own, underpinning and supporting the drive to improved governance. The problem lies, advocates would argue, in the plethora of contradictory and un-implementable laws, as much as in the lack of will to implement them. Others are concerned that, if the elements of the bushmeat trade are to be legalized, enforcement will be unavailable and ineffective, leading to a greater free-for-all than exists even now. Positive experiences are rather limited, and largely restricted to high-potential areas in southern Africa which benefit from both a relatively stable tourist trade and the proximity of industrial economies, but, allowing for these qualifications, they point to the value of the legal reform route, when conditions are otherwise favourable. For example, Namibia's experience of a rights-based approach to community natural resource management (discussed by Vaughan and Long in Chapter 8) is an interesting one, and may well be replicable. Two other chapters in this section (Chapters 9 and 10) examine the interplay between institutions and practice in the contexts of the Congo Basin. The first (by Poulsen *et al.*) examines the PROGEPP project in the area of the Nouablé–Ndoki National Park in Congo-Brazzaville, while the second (by Andrew Hurst) looks at this and two other experiences, in Cameroon and Gabon, in a comparative frame.

Conservation needs to be reconciled with development not just in practice and law but also in policy. This is an underdeveloped field and yet one that is immensely challenging at the present time. Development policy is increasingly orientated to national ownership and to supporting, rather than challenging, host government priorities. In relation to forestry in general, a shift in authority towards the producer states is almost always associated with marginalization within the national agenda, and this is likely to be equally true of the bushmeat subsector. There are several reasons for this. In the first instance, forestry is an unattractive sector for development assistance, being itself a major revenue earner. The donor interest tends to be more focused on the need to improve governance and the distribution of revenue than on generating additional financial support. Bushmeat is not a major revenue earner, and is unlikely ever to become one, but figures as part of a wider governance problem, to which producer governments are loath to draw attention. High international media interest but low likelihood of being able to make quick progress also make this a 'lose–lose' agenda for many bureaucrats, and not one that is likely to be career-enhancing for them.

Bringing bushmeat into national policy is clearly a major challenge, and one that needs to be confronted sooner rather than later. Brown (Chapter 7) considers the ways in which the issue might be brought into development policy. The development policy dimension is a crucial one, not just (in a negative frame) because of the problematic aspects of the shift away from the global to the national agenda, but also (in a more positive one) because of the new opportunities that this shift may also offer. These particularly concern the renewed interest in long-established but hitherto becalmed themes such as environmental and natural resource accounting. As the distinguished Ghanaian champion of the bushmeat issue, the late Emmanuel Asibey, forcefully advocated, there can be little hope of bringing the topic seriously into national policy as long as understandings of its value to producer economies remain so limited. A topic which is treated as illegitimate for public policy is hardly well placed to compete with those concerns (education, public health, etc.) that are universally regarded as central to it. An unparalleled opportunity exists at the present time to bring the bushmeat issue into public policy, both as a component of national wealth and environmental accounting, and also by virtue of its relationship to the other concerns – protein supply scenarios, ecosystem functioning, global climate change and the like – that are at the heart of current development debates.

This is clearly a field that will need to be given further attention if bushmeat is to figure at a level appropriate to the livelihood and other values documented in this book. In an effort to broaden the debate, a number of studies were included from other sectors or other regions (Part 3: Extrasectoral Influences and Models) – forestry for timber production, agriculture, non-timber forest products (NTFPs), non-tropical wildlife management. The interest here is in identifying those areas where natural resource management has been improved, and the lessons which can be learnt from these extrasectoral experiences can be fed into wildlife policy.

The attempt to view wildlife conservation within broader strategies of land use management, and to make links with agriculture and rural markets, presented in the COMACO case study from Zambia, provides a powerful set of principles of particular interest where human population densities are too high to sustain classical tourism-based approaches (Chapter 11). The broader picture on the plant NTFP trade offers interesting pointers to the ways in which supply chains can be managed in favour of the poor, including women (Chapter 12). Likewise, the improvements in the quality of resource management which can derive from small but radical shifts in public policy is well illustrated by the case study of barren-ground Caribou conservation in northern Canada (Chapter 14).

Experiences such as these are very close to the bushmeat trade, and are arguably of much greater interest to policy development than are more ambitious attempts to provide substitutes for the bushmeat trade. Though much promoted in the literature, these more radical ventures have shown little evidence of success to date. For example, the record on reconciling conservation and development agendas has not been strong. Integrated conservation and development projects (ICDPs) and attempts to develop alternative income-generating strategies to substitute for hunting have rarely had a major impact either on the condition of the resource or on the welfare of hunting communities. Tourism and sports hunting have also been promoted as alternatives but have rarely fulfilled their promise, particularly in the bushmeat heartlands. Lack of infrastructure and access, and high levels of insecurity for locals and travellers alike, have limited the appeal and added to the fragility of the local livelihoods. Except in a few celebrated cases, there does not appear to be much prospect of a radical transformation of producer economies by such means. Arguably, the attempt to reconcile conservation and development has produced solutions which have been to the satisfaction of neither constituency, and with little prospect of long-term sustainability. All things considered, the resolution of the 'bushmeat crisis' seems likely to lie more in control and management of the industry, and in governance of the forest sector, than in the search for alternatives disconnected from the existing economy.

Getting these findings into policy is not unproblematic, however. Early signs are that the new aid architecture is proving unconducive to natural resources in general, and to forestry in particular. Bird and Dickson (Chapter 13) consider the treatment of forestry themes (including non-timber products such as bushmeat) in poverty reduction strategies, and the evidence is not encouraging. Not only is the topic marginal to the policy debates, but also the means used to address issues of public concern (for example, participatory poverty assessments) are unlikely in many situations to give an accurate picture of it. Most of the harvest takes place on public lands in Central Africa, and poor villagers are often reluctant to draw attention to their use of such areas, given its quasi-illegality.

The final part of the book (Part 4: Regional Perspectives) looks beyond the West and Central African zone, which has been the main focus of interest, to the experience of other regions. The African focus is certainly justified in conservation terms. It is here that both the volumes of bushmeat extracted and the conservation threat are greatest. However, South American biodiversity may become threatened in a similar fashion in coming decades, though it is arguable that the bushmeat trade will be only a minor element in this, given the much greater production of domestic meat in Latin America (Part 4 Introduction) and the greater impact of deforestation and land conversion on animal biodiversity.

The situation in Asia is more directly instructive, as Bennett summarizes (Chapter 15), for the situation on this continent warns in many ways of the dangers of inaction, as human populations and demand increase. While South-east Asia may have historically been less dependent on bushmeat than Africa in aggregate terms, wild animal meat is still a highly valued commodity for many Asian forest dwellers. It is also highly valued by Asian urban consumers, who prize its combined food and medicinal properties – and in this context the rapidly expanding Chinese market looms large. Only in a few areas (Sarawak in Malaysia, for example) has there been an attempt to control the off-take, outside protected areas, and this offers an example of effective wildlife management – but there are conspicuously few others.

Conclusion

These are, then, interesting times for the student of the wild meat trade. New communications, technologies and patterns of trade have brought the peoples of the different continents together in ways that could never have been imagined even a few decades ago. The divergent values that wildlife represent exemplify these often conflicting cultural values, and raise important questions about the ways in which wide-ranging priorities and interests might be reconciled. The starting point must be to better understand the dynamics of the wildlife populations themselves, and the part they play in the human societies which exist in closest proximity to them. With a better understanding of both these dimensions, national and international decision-makers will be best placed to intervene to sustain positive processes of change. This volume is intended as a contribution to this cause.

⬭ **PART 1** ⬭

Bushmeat: Markets and Households

Glyn Davies and John G. Robinson

In this section we focus on the importance of bushmeat to human livelihoods, using various case studies from West, Central and southern Africa. In a study in Cameroon, it was found that, although cacao and coffee were the main sources of individual income, bushmeat was important for village hunters as a means of obtaining cash during the 'hungry season', when crops and food are scarce (Chapter 4). In neighbouring Equatorial Guinea, a study indicated that income from bushmeat hunting was similarly low overall, certainly much lower than construction work currently offers during the country's oil boom (Chapter 5). However, there are instances when sustained incomes can be achieved (Davies, 2002; Elliott, 2002), as noted for professional hunters in Ghana (Chapter 2) and Zambia (Chapter 6). Nevertheless, we still have a poor understanding of the contribution bushmeat makes compared with other items in individuals' food economies and annual income – especially during 'lean periods', or following economic or social shocks.

Other chapters in this section focus on markets and household consumption. A large number of different animal species are eaten, gifted, loaned or traded as bushmeat. Many are mammals (from the large elephant to the tiny elephant shrew), but birds, reptiles, snails and insects are regionally important. There are dramatic regional differences in the relative proportions of taxa present in markets (Chapter 3). Considering just three groups – ungulates (mostly duikers, other antelopes and pigs), primates (rarely including great apes) and rodents (mostly porcupines and cane rats) – a comparison of different markets (Table 1) shows that primates may account for between 1% and 50% of the carcasses in West African markets, and ungulates for between 38% and 88% in Central African markets. In an urban market in Ghana, snails and rodents predominate (Cowlishaw *et al.*, 2004). This variation derives from complex interactions between various ecological, cultural, social and economic factors.

Ecology defines source faunas, which are different between savannas and forests. Savannas tend to support much higher standing mammalian biomass, and a high percentage of species are ungulates (Robinson and Bennett, 2004). As the case study from Zambia reminds us, savanna species

Table 1 **Percentage contribution of animal taxa to total animal sales in bushmeat markets across Africa.**

Region/country	Ungulates	Primates	Rodents	Other species
West Africa				
Sierra Leone*	29.6	50.5	11.4	8.5
Ghana[†]	6.5	1.1	79.7	12.8
Ghana[‡]	7.8	0	81.0	11.1
Ghana[§]	65.7	1.2	31.1	1.8
Ivory Coast[§]	22.3	2.8	66.4	8.5
Nigeria[¶]	18.9	16.5	61.2	3.4
Central Africa				
Equatorial Guinea (1991)**	36.7	25.7	37.6	0
Equatorial Guinea (1996)**	38.1	5.2	56.7	0
Equatorial Guinea[††]	35.0	20.8	18.2	25.9
Gabon[‡‡]	58.0	19.0	14.0	9.0
Congo[§§]	70.0	17.0	9.0	4.0
CAR[¶¶]	77–86	0	11–12	2–12
Cameroon***	88.0	3.0	5.0	4.0

*Chapter 2, this volume.
[†]Asibey (1974).
[‡]Cowlishaw *et al.* (2004).
[§]Hofman *et al.* (1999).
[¶]Anadu *et al.* (1988).
**Fa *et al.* (2000).
[††]Kümpel (2006).
[‡‡]Lahm (1994), in Wilkie and Carpenter (1999).
[§§]Delvingt (1997).
[¶¶]Noss (1995).
***Dethier (1995).

involved in the bushmeat trade tend to be antelopes (Chapter 6). Relatively undisturbed forests support many large-bodied mammals of conservation concern – great apes, elephants, bongos, forest buffalo – which tend to be rare, and can be eliminated by even low levels of hunting and trapping. Moreover, many middle-sized mammals, such as forest monkeys, which can occur at high densities in relatively unhunted forest, can be quickly depleted because their reproductive rates can be quite low (Robinson and Bennett, 2000). At the other end of the habitat spectrum, bushmeat from heavily logged forests tend to be small-bodied, resilient species whose biology has allowed them to withstand human hunting, sometimes for centuries. Many of these species also thrive in farmbush areas and are commonly crop pests – cane rats, bushbuck, brush-tailed porcupine and blue duiker. The key point to appreciate is that it is not just forests, but also the vast areas of farmbush, that supply bushmeat to many people, as evidenced from cases in Sierra Leone (Chapter 1) and Ghana (Chapter 2).

Culture also constrains harvests. For instance, it is not only geneticists who have noticed the strong similarity between great apes and humans, but many rural communities have rules, tales and understandings about great apes. Whereas some groups preferentially hunt great apes (Starkey, 2004), in other areas there is a taboo against eating great apes, as noted for Sierra Leone (Chapter 1), parts of Equatorial Guinea (Chapter 5) and the eastern Democratic Republic of Congo (Kalpers *et al.*, 2003). Only in extreme economic hardship do these taboos break down, until which time they can help underpin local management systems, as can Muslim taboos on consumption of monkey meat and pork.

Hunting technology, to a large extent, determines the species selected. The techniques used to obtain bushmeat include traditional methods such as bows and arrows, net hunts (with people or dogs driving wildlife into nets), spears and the use of twine traps. More modern methods involve wire snares to catch antelopes and pigs, but the occasional carnivore, great ape or buffalo is also taken (Chapter 3). Guns are used to kill diurnal species such as monkeys, and at night, with the aid of headlights, to kill antelopes and other ground dwellers. In a case in Equatorial Guinea, a reduction in the gun hunting of monkeys followed a presidential ban on firearms in the 1970s (Butynski and Koster, 1994) but recently primate hunting has increased again as guns have once more become more widely available and affordable (Kümpel, 2006).

The nature and identity of hunters affect what technologies are used, what source faunas are exploited and thus what species are harvested. At one extreme, there are full-time hunters, who opportunistically enter newly accessible areas, such as those created by logging concessions. Often they are exploiting relatively undisturbed forests, tend to hunt from temporary camps and seek to capture all marketable species within an area, including endangered and protected species. They tend not to be indigenous to the area, and are often supplying urban markets. They may hunt alone, such as those working on the edge of Monte Alén National Park in Equatorial Guinea (Chapter 5), or in groups that walk into the depths of the forest, such as those living in the Dja National Park, Cameroon (Chapter 4). At the other extreme, there are people who make their living from other activities, such as small-scale farming. They hunt on a part-time basis, setting traps at field edges, and in the farmbush areas around their fields and villages, getting a meal and reducing crop losses. They may also patrol their coffee and cacao crops with shotguns, killing guenon monkeys and squirrels (Chapter 1).

Social networks, and the chain of custody this creates, introduce another filter into what gets to market. A number of studies in this section explore how bushmeat moves to market, from hunters and trappers, to transporters and traders, to wholesalers, 'chop bar' owners and individual households. The relationships between these groups can change with political circumstance (de Merode *et al.*, 2004) and, as a result, the extent of benefits and their distribution also varies.

The policy implications of this situation are several. First, there are many different social categories of people who depend to a lesser or greater extent on the hunting and sale of bushmeat. In many areas, raising domestic livestock is difficult, and people depend on wild meat for their animal protein. Generalized prescriptions such as 'ban the bushmeat trade' are both impracticable and unfeasible in these circumstances.

Second, in many areas, much of the hunting involves species not threatened or endangered, often in areas already heavily disturbed and degraded, and the bushmeat trade is of limited conservation importance. Conservation resources are probably better focused elsewhere (Bennett *et al.*, 2007).

Third, there is frequently a linkage between the bushmeat trade and areas of forests being exploited for timber. Many of the forests in the Congo Basin still contain large-bodied mammals of conservation importance, and there is an urgent need to engage in policy dialogue on sustainable forest management, and trading in certified timber, to ensure that wildlife management is addressed in production forests.

Fourth, hunting of wildlife must be controlled in parks and protected areas. All conservation efforts in protected areas and production forests rely on a supportive policy environment – not conflicting policies on mining, roads and agriculture.

Ultimately, all management of the bushmeat trade will rely on transparent and accountable governance, so that legitimate conservation and livelihoods concerns can be addressed in different ways in different areas.

Hunting and Trapping in Gola Forests, South-eastern Sierra Leone: Bushmeat from Farm, Fallow and Forest

Glyn Davies, Björn Schulte-Herbrüggen, Noëlle F. Kümpel and Samantha Mendelson

Introduction

Bushmeat harvesting is an important livelihood activity in West and Central Africa. It is a source of animal protein (e.g. Anstey, 1991; Chardonnet *et al.*, 1995; Elliott, 2002; Fa *et al.*, 2002a; this volume), and a source of income for many rural economies with few alternatives (Anstey, 1991; Juste *et al.*, 1995; de Merode *et al.*, 2004; Cowlishaw *et al.*, 2005a). It can also be an important part of crop pest management (Davies, 1990; Naughton-Treves *et al.*, 2003). The term 'bushmeat' encompasses all wild species eaten for meat, from insects and mice to gorillas and elephants, and this 'multispecies' nature of the bushmeat trade complicates the understanding of livelihood, conservation and management issues.

The Gola Forest Project focused on mammals weighing more than 3 kg, and sought to investigate the importance of different habitats as supply areas for bushmeat by:

(a) comparing primate and ungulate populations in unlogged forest, logged forest and farmbush; and
(b) examining bushmeat harvest, trade and consumption at three different levels: farmer households on the edge of the forest; hunters operating in the forest and farmlands; and the bushmeat market in Kenema town, 30 km from the forest.

Study sites

The Gola Forest Reserves (Figure 1.1) cover 784 km² and were gazetted in 1926 and 1930 as timber production reserves. They are the last substantial areas of tropical rain forests in the Republic of Sierra Leone (Davies, 1987), and are part of the Upper Guinean forest biodiversity hotspot (Oates, 1998). The Reserves

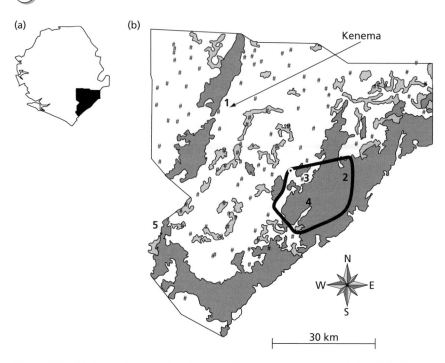

Figure 1.1   (a) **Sierra Leone, showing the Kenema region around the Gola Forest Reserves (# indicates village location). (b) The approximate catchment supplying bushmeat to Kenema market, and the 'hunting zone' of the Lalehun hunters (heavy black line). The vegetation types are: forest (dark grey), farm-fallow matrix (white) and heavily degraded forests strips (light grey). From FAO (1979). Kenema and the survey sites are indicated: 1, Kondebotihun; 2, Mogbai; 3, Lalehun; 4, Koyema; 5, Tiwai island.**

lie between 100 and 300 m above sea level, with maximum temperatures reaching 40°C during the dry season and average minimum temperatures of 22.5°C. Taking meteorological data from Tiwai Island, on the western edge of the Gola Forests, the annual rainfall averages 3,000 mm with a pronounced rainy season from May to October and a dry season from December to March (Oates *et al.*, 1990).

Kenema town is the main urban centre in the region, and the headquarters of Kenema District, while the focus of our village surveys was in Lalehun village, a forest-edge village that is on the main timber track leading into the Gola Forest Reserves.

Botanical surveys

The Gola Forest Reserves are on the eastern flank of an agricultural landscape that comprises a mosaic of farms, farm-fallow, cacao/coffee plantations and forest patches, and the forest in the reserves comprises long-established forest habitat that grew back from farmlands and secondary forest when wars depopulated the area in the mid-nineteenth century (Unwin, 1909). Botanical surveys were carried out during this study in two forest habitats – unlogged and logged forest – and two farmbush plots. Trees were measured in plots or subplots of large botanical quadrants in two categories: big trees (girth > 100 cm) and small trees (girth 30–100 cm); tree stumps were also measured (Table 1.1). These data were compared with data from similar-sized botanical plots in 60-year-old closed forest (Oates *et al.*, 1990) on Tiwai studied previously (Davies, 1987).

Looking at the top five species in the botanical samples there is the expected variation in tree species abundance (Table 1.1). However, there is also considerable overlap in terms of species present, confirming that all represent a similar forest formation, including Tiwai (despite having regenerated after being cleared and burned some 60 years previously). Small trees sampled at these sites reflected understorey species as well as young forms of larger species. In the farmbush sites, colonizing species such as *Macaranga* and *Musanga* predominated at Tiwai and Lalehun, but were less dominant at Kondebotihun, where frequent pole cutting, firewood collection and farming resulted in a thicket of coppice stems of other species.

On Tiwai, although there were many more trees per hectare in the farmbush, the lack of any large trees meant that the stem basal area in farmbush is approximately 50% lower than in mature forest (Fimbel, 1994). Farmbush seldom has trees reaching more than 10 m high, in contrast to forest habitats, which are multilayered and include trees over 45 m tall.

Wildlife surveys

In the two forest sites (Mogbai and Koyema), a standardized survey grid was used to census primates: 1 km × 500 m grid with parallel transects 100 m apart. The perimeter of the grid was walked slowly (at approximately 0.5 km/h) for five consecutive days, in alternating directions each day, in four or five different months (Table 1.2), and all primate group sightings and calls were mapped. In addition, 5 days was spent at both sites carrying out 'sweep surveys' (Whitesides *et al.*, 1988), which use similar information but gathered simultaneously by a team of surveyors moving along parallel transects to get more accurate group density estimates. The density of individuals was then determined by multiplying group densities by the average number of individuals per group, using data from long-term studies on Tiwai (Whitesides *et al.*, 1988). In the farmbush sites (Lalehun and Kondebotihun), the dense thickets meant that only a rectangular transect could be walked, with all sightings and calls being mapped, without the intervening grid.

Table 1.1 **Characteristics of habitat, hunting and tree species at three forest and three farmbush survey sites.**

Survey site	Habitat and land use (sample area; tree density; no. of species)	Hunting/trapping	Five commonest large tree species (no. of stems > 100 cm gbh/ha)	Five commonest small tree species (no. of stems > 30 cm gbh/ha)
Tiwai forest*	Mature secondary forest: (>120 cm gbh: 8 ha, 50.2 trees/ha, 33 spp.; >30 cm gbh: 1 ha, 301 trees/ha, 56 spp.)	Very little gun hunting (local hunting ban); some trapping to protect farms; no permanent habitation	*Pentaclethra macrophylla* (7.9), *Parinari excelsa* (5.8), *Sarcoglottis gabonensis* (2.6), *Uapaca guineensis* (2.5), *Kaoue stapfiana* (2.6)	*Diospyros thomasi* (41.8), *Kaoue stapfiana* (20.9), *Xylopia acutiflora* (24.5), *Xylopia elliotti* (20.9), *Gilbertiodendron* spp. (19.9)
Mogbai forest	Unlogged forest (>100 cm gbh: 9.8 ha, 47 trees/ha, 69 spp.; >30 cm gbh: 1.96 ha, 351 trees/ha, 95 spp.)	Low gun hunting/trapping pressure (3–4 h walk to nearest village); no snares or cartridges found; local hunters reported infrequent use	*Heritiera utilis* (3.9), *Didelotia idae* (3.7), *Pentaclethra macrophylla* (2.3), *Kaoue stapfiana* (2.3), *Sarcoglottis gabonensis* (1.7)	*Hariteria utile* (68), *Kaoue stapfiana* (54.5), *Diospyros sanza-minka* (29), *Uapaca guineensis* (23.5), *Didelotia idae* (20.5)
Koyema forest	Selectively logged forest (> 100 cm gbh: 10 ha, 31.6 trees/ha, 57 spp.; stumps 7.9/ha; >30 cm gbh: 2 ha, 648 trees/ha, 139 spp., 5 stumps/ha)	Moderate to high hunting in recent past; old hunters' camp and drying racks nearby (intensive hunting September 1986 to April, 1987)	*Pentaclethra macrophylla* (13), *Uapaca guineensis* (14.6), *Funtumia africana* (6.5), *Parinari excelsa* (3.4), *Piptadeniastrum africanum* (3);	*Funtumia africana* (48), *Anthonotha macrophylla* (29), *Myrianthus libericus* (24), *Pentaclethra macrophylla* (19), *Uapaca guineensis* (17)

Tiwai[†] farmbush	Farmbush fallow, 5–12 years after cultivation	Very little gun hunting (local hunting ban); some trapping to protect farms; no permanent habitation	n/a	Ochthocosmus africanus (519), Musanga cercropiodes (350), Samanea dinklagei (331), Macaranga barteri (214), Daniellia thurifera (192)
Lalehun farmbush	Cleared, felled and burned 6–8 years previously, first or second time (>100 cm gbh: 0.38 ha, 26/ha, 10 spp; >30 cm gbh: 0.38 ha, 1,950 trees/ha, 70 spp.)	Moderate to high gun hunting pressure; commercial and subsistence trapping around the edge of farms	Cynometra leonensis (5), Dialium guinensis (5)	Musanga cecropiodes (503), Macaranga heudelotti (312), Funtumia africana (140), Ficus capensis (50), Macaranga barteri (41)
Kondebotihun farmbush	Patchwork of 3- to 6-year-old fallow land, farmed several times; repeated fuelwood collection (>30 cm gbh: 0.42 ha, 3,620 trees/ha, 56 spp., 3,620 stumps/ha)	Very strong gun hunting and trapping pressure (only 1 km from Kenema town)	n/a	Xylopia aethiopica, Anthostema senegalense and Cathormion altissimum

gbh, girth at breast height; n/a, not analysed.
*120 cm gbh measured on Tiwai (Davies, 1987).
†Fimbel (1994).

Table 1.2   **Site specific details about wildlife census at four sites (see Figure 1.1b for a map of the sites).**

Survey site	Survey months (1988 to 1990)	No. of survey days	Distance walked (km)
Mogbai	Nov, Mar, Apr, May, Dec	25	60.2
Koyema	Dec, Jan, Feb, June, Nov	26	71.7
Lalehun	Sept, Oct, Dec, Jan, Nov	20	50.0
Kondebotihun	Oct, Nov, Dec, Jan	15	63.0

All sites were surveyed in the drier months as the main rainy season (June to August, 1989) was avoided because poor weather reduced visibility and calling frequency (Table 1.2). On Tiwai, primate population densities were estimated during long-term studies and using sweep samples in the forest (Whitesides *et al.*, 1988), and using transect surveys in the farmbush (Fimbel, 1994).

A limitation of the grid survey method is that primate species whose annual home ranges exceed the 1 km × 500 m grid are seldom recorded within the sample zone. This applies to sooty mangabeys (*Cercocebus atys*) and the rare chimpanzee (*Pan troglodytes*), both of which generally travel long distances along the ground and have large home ranges. Both species were recorded in (or very near) the survey grids, but realistic population density estimates were possible only from long-term studies of mangabeys in Tiwai farmbush and forest, and not for chimpanzees at any site.

For the duikers, transect surveys around the grid produced over 35 sightings per site of the commonest species, Maxwell's duiker (*Cephalophus maxwelli*). However, it is clear that transect methods for estimating duikers are very unreliable (Davies *et al.*, 2001; Newing, 2001), and the situation was even less satisfactory for the large but cryptic rodents, such as cane rat, *Thryonomys swinderianus*, and brush-tailed porcupine, *Atherurus africanus*, which were not sampled effectively at all despite being an important component of the bushmeat trade.

Bushmeat sources

Within the three forest sites, Tiwai had the least hunting pressure and the highest primate biomass (Figure 1.2). This was noteworthy because the Tiwai forest was not primary, but had regenerated to forest following farm clearance some 60 years earlier. Primate populations declined markedly as hunting pressure increased, with Mogbai being excellent primary forest habitat. Red colobus numbers accounted for much of the difference between forest sites, being highest on Tiwai and lowest at Koyema, where a hunter reported killing about 500 monkeys between September 1986 and April 1987. The Diana monkeys, *Cercopithecus diana*, appear less vulnerable than red colobus to

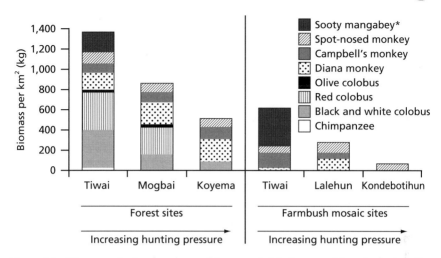

Figure 1.2 **Biomass of primate communities recorded in forest and farmbush sites. No biomass estimates were included for species that were recorded at a site but with too few sightings to estimate their density. *Surveyed only on Tiwai.**

hunting pressure, but their numbers did decline as forest gave way to farmbush – those in the farmbush at Lalehun were found only in an area adjacent to a clump of large trees.

In the case of Campbell's (*C. campbelli*) and spot-nosed monkeys (*C. petaurista*), numbers seemed remarkably similar in logged forest, unlogged forest and farmbush, irrespective of hunting pressure – their smaller body size makes it easier for them to move through farmbush thickets and evade hunters. Only in Kondebotihun, where hunting pressure was very high and farmbush quality very poor (there were very few trees), did their numbers drop dramatically. Using standard group sizes and body weights (Appendix 1.1), the total estimated biomass of the primate community was higher in forest than in farmbush (Figure 1.2), but increasing hunting pressure in either forest sites or farmbush sites was associated with decreasing primate biomass.

Looking separately at the sooty mangabeys on Tiwai, their numbers showed an increase in farmbush compared with forest – which was the opposite of the more arboreal species.

Duikers were present at all sites, but densities could be estimated only at the forest sites. If the transect sightings give a relative index of population density, then habitat change from logging did not appear to reduce duiker densities (Davies *et al.*, 2001), even with some increased hunting pressure at Koyema. In the farmbush, however, only Maxwell's duikers were recorded, and then only rarely.

Taking these results together, forest habitats clearly support a wider range of primate and duiker species than farmbush, and forests support larger populations of large-bodied colobine species. Farmbush and forest supports similar populations of the two smaller guenons, Campbell's and spot-nosed monkeys, as well as sooty mangabeys, and probably moderate to high numbers of Maxwell's duiker (Davies *et al.*, 2001). To this can be added bushbuck, *Tragelaphus scriptus*, and red river hog, *Potamochoerus porcus*, which occur both in forests and in farmbush (Davies, 1987). Of the large rodents, cane rats are abundant in farmlands and brush-tailed porcupines are abundant in forest clearings/margins.

Crop pests

The main agricultural activity in the Gola region is upland rice growing, with associated intercropping (Richards, 1985). Fields often have a perimeter fence into which small gaps are made for setting traps. Surveys of 135 farms in the vicinity of 45 villages in the southern province (adjacent to the Kenema District study area) assessed rice damage in 1-m-diameter plots spaced at 10-m intervals along two transects that crossed the centre of each field at right angles. The results showed that rats damaged 21% of the plots and cane rats damaged 24%, equivalent to 1% and 6% of the rice crop respectively – with peripheral plots showing greatest damage (Davies, 1990). In response, preliminary interviews with 12 Lalehun farmers indicated that they set between 10 and 30 fence-line snares, and caught between 0 and 19 cane rats each in one growing season, although more thorough surveys are needed to get an accurate estimate.

Farmers also grow cacao and coffee as cash crops; within Lalehun and two adjacent villages 58% of 1,670 records of income generation related to tree crop harvesting (Davies and Richards, 1991). Cacao and coffee trees were generally grown under shade trees in 'thinned forest', and spot-nosed and Campbell's monkeys, as well as mangabeys, were reported to be severe pests during the fruiting season. Squirrels also do substantial damage (Taylor, 1961). In the 1940s and 1950s, monkey damage in southern and eastern Sierra Leone was estimated at 20–25% loss of food and cash crops (T. S. Jones, 1998). In some cases, tree traps are set to reduce this damage, but most pest control is done by hunters shooting monkeys in the early morning and late evening.

Household and market surveys

A comparison of the 1963, 1974 and 1985 censuses in Kenema District indicates a human population increase of 48% over the period, with 337,055 people recorded in 1985. Looking at a subset of those chiefdoms on the edge of Gola North Forest Reserves, there are about 50,000 people, with a population density of approximately 30 persons/km^2 (Davies and Richards, 1991). Socioeconomic surveys led by Paul Richards focused on 12 villages, comprising 3,202 people from the area adjacent to Gola North Forest Reserve (equivalent to some 25%

of all villagers living adjacent to Gola North Forest Reserves), with particular attention being focused in Lalehun and two adjacent hamlets.

Lalehun household bushmeat consumption

To examine bushmeat consumption patterns, 35 women from Lalehun were interviewed, based on a 7-day recall in three sample periods: the height of the hungry season (August–September, 1989); the harvest season (November–December, 1989); and the end of the dry season (April, 1990). Interviewees stated the number of times per 7 days they had used a particular item when cooking (Davies and Richards, 1991).

The household survey revealed that ,whereas fish, reptiles, crustaceans and amphibians were included in 60% of all meals, mammals were only recorded in 37% (n=722). In terms of the mammal species being consumed (n=269), two antelope species (bushbuck and Maxwell's duiker) accounted for 33% of records, whereas three rodent species (cane rat, brush-tailed porcupine and giant rat) accounted for 38%. Primates accounted for only 6% of mammal records consumed in the households. Over 90% of animals consumed were 'generalists' found in both farmbush and forest habitats, with only two records of rare species being consumed (chimpanzee and bongo, *T. euryceros*). Furthermore, cane rats were eaten more frequently during the late rainy, rice-growing (and hungry) season, whereas duikers were more important in the dry season.

Villagers were also asked if there were any foods they would not eat. Of 127 records of taboo animals, chimpanzees were the most commonly cited (18.1%), followed by monkeys (all species) at 10.1% and snakes (9.4%).

Lalehun hunters and Kenema market

Hunting, diamond digging and commercial cassava growing were recorded as the most prevalent income-generating activities for men around the Gola Forests (Davies and Richards, 1991). In Lalehun there were 11 hunters with guns, and four local field assistants recorded the dates of hunting trips, the hunter's name, the species killed and the hunting technique. This was done continuously from May 1989 until February 1990 (amounting to 246 days surveyed). Care was taken not to double count any hunter's bag on a given day, but it is often difficult to record what meat might have been consumed during a hunting trip, or what endangered animals were secretly killed (Hartley, 1993).

It is conspicuous that the Lalehun hunters' bag records show that 66% of traded items were shot, compared with 77% of the food items consumed in households which were trapped (Davies and Richards, 1991), indicating that many cane rats and duikers were being trapped by farmers, on farm, in addition to the hunters' bag. Furthermore, the Lalehun hunters most frequently exploited logged forest (57%) – the nearest and most accessible forest habitat

– followed by farmbush (33%) and more distant primary forest (7%), with farms and cacao/coffee plantations accounting for the rest (Figure 1.3).

Bushmeat entering the Kenema market was recorded by a Sierra Leonean field assistant over the same 10-month period (May 1989 to February, 1990) for a minimum of seven days per month (different days: Monday to Friday; $n = 109$ days). This was done between 08:00 and 12:00, when most *podapoda* minibuses arrived carrying bushmeat from rural and forest areas. Detailed information was collected on species and price (see Figure 1.6).

In order to get an impression of the monthly variation between what was hunted at the forest edge in Lalehun (Figure 1.4a) and what was traded in the urban centre in Kenema some 30 km away, supplied from forest and farmbush areas (Figure 1.4b), a comparison was made between the two sites.

The two sets of data were analysed (Kolmogorov–Smirnov tests, two-tailed; Figure 1.5), and show that the proportions of four taxonomic groups in the Lalehun hunters' bags and Kenema market sales were significantly different ($\chi^2 = 446.18$, d.f. $= 3$, $P < 0.001$). Ungulates constituted 66% of all animals in the Lalehun hunters' bags compared with 30% in Kenema, and primates only 18% in the Lalehun bag compared with 50% of Kenema sales (Figures 1.4 and 1.5). Rodents constituted a small proportion of market sales and were rarely recorded in hunters' bags, despite their preponderance in household consumption. The implication here is that the Lalehun forest-edge hunters sold primates, preferring to keep antelopes for local consumption. This corresponds to the low levels of primate consumption in Lalehun households, also noted above. Rodents, similarly, were consumed preferentially to primates in Lalehun households.

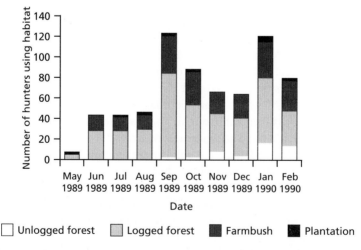

Figure 1.3 **Number of times different habitats were used by hunters with guns.**

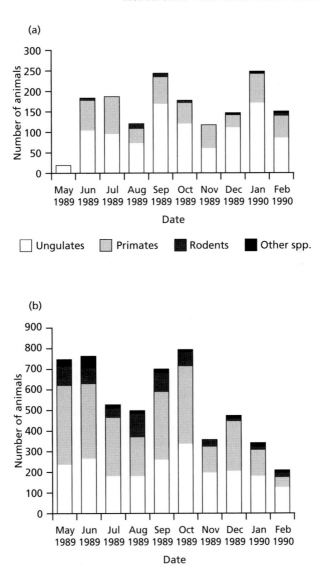

Figure 1.4 **Number of animals from different taxonomic groups: (a) making up hunters'
bag at Lalehun and (b) entering Kenema market from May 1989 to February 1990.**

The proportion of generalist, forest-dependent and rare species also varied. Three species, Campbell's monkey, spot-nosed monkey and Maxwell's duiker, accounted for 60% of all Kenema market sales. The proximity of forest to the Lalehun hunters meant that the hunters' bag had a higher proportion of forest-dependent species: 23% compared with 6.3% in Kenema market ($\chi^2 = 232.20$, d.f. $= 2$, $P < 0.001$). Rare species were recorded for only 1% of all market or Lalehun hunters' kills.

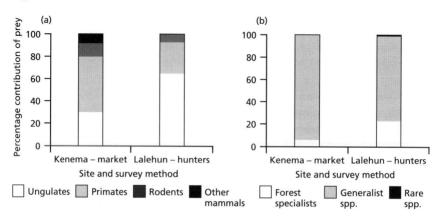

Figure 1.5 **The relative contribution of different prey groups by (a) taxonomic groups and (b) ecological guilds: rare (zebra duiker, chimpanzee, bongo), forest dependent (bay duiker, black-and-white colobus, red colobus, olive colobus, diana monkey) and generalist species (the remaining mammal species) (after Davies, 1987).**

Discussion

This study shows that bushmeat is trapped and hunted in farm, farmbush, plantation and forest habitats in eastern Sierra Leone. Forests support a wider range of species, including many forest-dependent and rare species, at a higher biomass. Diversity and density are reduced as a result of hunting and trapping, despite many species being able to survive vegetation changes, as indicated by the Tiwai results (Davies *et al.*, 2001; Plumptre and Grieser Johns, 2001). Moreover, larger species, which are particularly sought by hunters, tend to be eliminated rapidly, as shown for red colobus.

Religious taboos have a strong influence on species selection in Lalehun and Kenema. The most conspicuous example is against eating chimpanzees specifically, and primates in general. An aversion to eating primates was recorded in the 1940s and 1950s (T. S. Jones, 1998), and possibly resulted from the strong influence of Muslim and Christian missionaries in Sierra Leonean society during the early twentieth century. Their harvest was virtually absent until the government 'monkey drives' of the 1940s and 1950s, during which 243,763 monkeys were killed, and Liberian hunters and traders were encouraged to make use of the glut of primate bushmeat (T. S. Jones, 1998). The high levels of monkey consumption and trade noted during this study can therefore be attributed to a change in human behaviour brought on by economic necessity and a lack of domestic meat for poor urban and rural consumers. Primates are not a delicacy for Sierra Leonean urban consumers, and even with this change in diet a strong aversion to eating chimpanzee meat was still apparent.

Taste and price are two other important factors that can influence bushmeat choice. In Lalehun, mangabeys were less preferred for trading because their oily flesh was difficult to dry and transport, whereas red colobus was described by many as 'having a bad smell' when fresh, but it dries well and was therefore a favoured species for trade. As a result, hunters often target red colobus and eliminate them from suitable habitats.

Prices of bushmeat per kilogram in Kenema market were slightly lower than for beef (Figure 1.6), although basically similar until July 1989, when the price of beef rocketed following a collapse in petrol supplies and a further downturn in the economy, both affecting beef supply. At present, bushmeat is substantially cheaper than beef, further increasing sales of the former.

Finally, availability of bushmeat from different source habitats will determine bushmeat supplies to markets – as some species become scarcer in the forests or farmbush, so other species will come to predominate in markets. This is perhaps best shown by comparing the Kenema market with the urban Takoradi market in Ghana, which is long established and has few remaining forest areas to provide bushmeat (Table 1.3). The predominance of primates

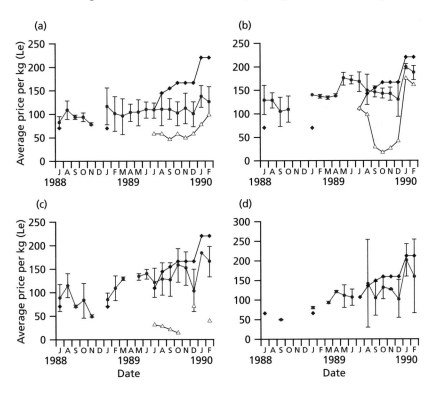

Figure 1.6 **Prices of (a) ungulates, (b) primates, (c) rodents and (d) other species (pale grey dots) compared with beef (black diamonds) by month over the study period. White triangles correspond to the prices paid to Lalehun hunters.**

in Kenema is very high, yet they are almost absent in Takoradi market, and the body size of species in the Takoradi market is generally smaller than in Kenema. This in turn means that the species found in the more intensively hunted region of southern Ghana are small-bodied and have the highest intrinsic rate of reproduction (r_{max}).

Conclusions

Bushmeat is clearly important to people's livelihoods in south-eastern Sierra Leone: as a food source, a source of income, and a way to reduce crop losses from mammal pests. Most of the trade appears to be underpinned by farmbush species and, under these circumstances, a general ban on the bushmeat trade would be inappropriate given the negative impact on human livelihoods, which would outweigh the small conservation gains. However, where the

Table 1.3 **The most frequently recorded mammal species in the Kenema and Takoradi market. The species are ordered according to their estimated intrinsic rate of natural increase (r_{max}).**

		r_{max}	Kenema	Takoradi*
Multiple spp.	Ground squirrels	1.47	–	Present
Anomalurus spp.	Flying squirrels	1.09	–	Present
Cricetomys emini	Giant rat	1.09	–	Present
Atherurus africanus	Brush-tailed porcupine	0.87	Present	Present
Neotragus pygmaeus	Royal antelope	0.69	Present	Present
Thryonomys swinderianus	Cutting grass	0.68	Present	Present
Civettictis civetta	African civet	0.51	Present	–
Cephalophus maxwelli	Maxwell's duiker	0.46	Present	Present
Cephalophus zebra	Zebra duiker	0.46	Present	–
Cephalophus dorsalis	Bay duiker	0.43	Present	Present
Cephalophus niger	Black duiker	0.42	Present	Present
Tragelaphus scriptus	Bushbuck	0.37	Present	Present
Cercopithecus petaurista	Lesser spot-nosed monkey	0.19	Present	–
Cercopithecus campbelli	Campbell's monkey	0.19	Present	–
Cercopithecus diana	Diana monkey	0.17	Present	–
Cercocebus atys	Sooty mangabey	0.14	Present	–
Piliocolobus spp.	Red colobus	0.14	Present	–
Colobus polykomos	Black and white colobus	0.14	Present	–

– = Species is not present.
*Cowlishaw *et al.* (2004).

bushmeat trade does have a negative impact on forest-dependent and rare species, regulation to exclude these species from the trade could achieve conservation goals without threatening livelihoods. Certainly, preventing hunting and trapping of bushmeat in protected areas (e.g. national parks) and focusing regulation on threatened species are essential to achieve conservation outcomes.

Acknowledgements

The biological and market surveys were led by Glyn Davies, and the socioeconomic surveys were led by Paul Richards. We were funded by ESCOR of the UK Overseas Development Administration, through a grant to University College London, which was implemented in collaboration with the Forestry Division of the Ministry of Agriculture, Natural Resources and Forestry and Njala University College in Sierra Leone. The wildlife and bushmeat data collection was assisted by Mohamed Bakarr, Sam Edwards, Helen Newing, Ranger Williams, Bockarie Gbandwa, Macfoy, Amara; Agatha Iye Kallon, Mama Zombu, and botanical surveys were led by S. K. Samai and V. K. Kallon. Data analysis was conducted by Björn Schulte-Herbrüggen, Noëlle Kümpel and Samantha Mendelson.

Appendix 1.1   Density (individuals/km²) of primates and duikers in forest and farmbush in east Sierra Leone.

Site	Forest			Tiwai‡	Farmbush	
	Tiwai*	Mogbai†	Koyema†		Lalehun†	Kondebotihun†
Hunting pressure	None	Low	Medium to high	None	Medium to high	High
Species						
Chimpanzee (Pan troglodytes)	1.0	+	+	+	+	+
Black and white colobus (Colobus polykomos)	51.5	22.5	13.5	(+)	–	–
Olive colobus (Procolobus verus)	7.2	7.5	–	(+)	–	–
Red colobus (Procolobus badius)	60.7	43.8	+	+	–	–
Sooty mangabey (Cercocebus atys)	39.0	+	+	73.5	+	–
Diana monkey (Cercopithecus Diana)	55.6	75.0	75.0	14.0	40.0	–
Campbell's monkey (Cercopithecus campbelli)	41.5	49.0	56.0	67.2	35.0	7.0
Spot-nosed monkey (Cercopithecus petaurista)	53.1	42.0	35.0	30.8	43.4	28.0
Duikers (Cephalophus maxwelli, C. dorsalis and C. zebra)	+	30	30	+	+	+
No. of primates recorded	8	8	7	8	5	3

Primates are arranged in decreasing order of body weight.
+ = the species was present and foraged at the site but scarcity of encounters prohibited density estimates; (+) = the species was not encountered foraging at the site but passing through; – = the species was not encountered; (–) = no data available.
*Oates et al. (1990).
†This study.
‡Fimbel (1994).

Appendix 1.2   **Percentage contribution of animal species to market sales and hunters' bags in Kenema and Lalehun .**

Species	Animals		Biomass	
	Kenema	Lalehun	Kenema	Lalehun
Ungulates				
Maxwell's duiker (*Cephalophus maxwelli*)	18.5	48.4	12.6	43.6
Bushbuck (*Tragelaphus scriptus*)	6.6	2.5	23.5	3.1
Red river hog (*Potamochoerus porcus*)	1.7	0.4	11.3	3.6
Water chevrotain (*Hyemoschus aquaticus*)	0.9	0.2	0.8	0.1
Bay duiker (*Cephalophus dorsalis*)	0.6	8.8	1.0	19.2
Black duiker (*Cephalophus niger*)	0.5	1.9	0.9	0.6
Buffalo (*Syncerus caffer*)	0.4	0	20.8	0
Royal antelope (*Neotragus pygmaeus*)	0.3	1.5	0.1	3.5
Zebra duiker (*Cephalophus zebra*)	0.1	0.7	0.1	1.4
Red duiker (*Cephalophus rufilatus*)	–	0.1	–	0.0
Primates				
Spot-nosed monkey (*Cercopithecus petaurista*)	17.6	5.1	4.5	2.5
Campbell's monkey (*Cercopithecus campbelli*)	14.6	4.5	5.5	3.9
Primates, unidentified multiple spp.	9.7	–	4.8	–
Sooty mangabey (*Cercocebus atys*)	3.0	2.1	2.0	2.1
Olive colobus (*Procolobus verus*)	2.8	0.3	1.0	0.2
Red colobus (*Procolobus badius*)	1.4	9.3	0.9	3.5
Black and white colobus (*Colobus polykomos*)	0.6	2.9	0.4	1.4
Diana monkey (*Cercopithecus Diana*)	0.4	1.8	0.1	1.4
Chimpanzee (*Pan troglodytes*)	0.4	0.1	1.0	0.3
Green monkey (*Cercopithecus aethiops*)	–	1.5	–	0.0
Cane rat (*Thryonomys swinderianus*)	10.9	2.2	6.1	0.9
Crested porcupine (*Hystrix cristata*)	–	3.8	–	8.3
Brush-tailed porcupine (*Atherurus africanus*)	0.5	0.4	0.1	0.1
Giant pouched rat (*Cricetomys emini*)	–	0.2	–	0.0
Ground squirrel, various spp.	–	0.1	–	0.0
Subtotal	*11.4*	*6.7*	*6.3*	*9.3*
Other mammals				
Civet/genet (*Viverra civetta/Genetta pardina*)	4.9	0.1	0.9	0.1
Pangolin (*Uromanis tetradactyla/P. tricuspis*)	–	0.4	–	0.1
Egyptian mongoose (*Herpestes ichneumon*)	–	0.3	–	0.1
Cusimanse (*Crossarchus obscurus*)	–	0.1	–	0.0

The data are expressed as number of animals and biomass per species. Biomass was calculated using adult body weights from Kingdon (1997).

– = No records of the species.

Livelihoods and Sustainability in a Bushmeat Commodity Chain in Ghana

Guy Cowlishaw, Samantha Mendelson and
J. Marcus Rowcliffe

Introduction

Bushmeat is a valuable natural resource that can play an important role in rural and urban livelihoods in tropical Africa. Bushmeat is not only a widely eaten food (e.g. Wilkie and Carpenter, 1999), it is also a market commodity that generates a cash income for those households that harvest and trade it (Dei, 1989; de Merode *et al.*, 2004). The value of bushmeat is reflected in national estimates of its domestic trade in West and Central Africa, which range from US$42 million to US$205 million (Davies, 2002). The growing evidence that the current trade in bushmeat is unsustainable is therefore cause for grave concern, as it poses a threat not only to the survival of those species that supply the trade, but also to the livelihoods of those people who depend upon it (e.g. Milner-Gulland *et al.*, 2003).

The development of a sustainable trade requires detailed knowledge of the biological limits of bushmeat harvesting, the livelihoods of its traders and the operation of its markets. The bushmeat trade in Ghana, West Africa, may provide valuable insights in this respect. Ghana is a densely populated country with an economy based primarily on smallholder agriculture (particularly cocoa) and with a long tradition of bushmeat trade and consumption (Grubb *et al.*, 1998). As long ago as the fifteenth century, De Marees recorded a trade in bushmeat over long distances (Clark, 1994), and, more recently, it has been suggested that the development of the cocoa industry in Ghana in the twentieth century would have been impossible without bushmeat's supporting role in smallholder livelihoods (Asibey, 1966; 1977).

The purpose of this chapter is therefore twofold: first, to provide an account of the bushmeat trade in Ghana, including the recently identified linkages between bushmeat and fisheries and, second, to provide a case study of the livelihoods and biological sustainability of the bushmeat trade in the city of Sekondi-Takoradi (hereafter Takoradi).

The bushmeat trade in Ghana

A national bushmeat survey in Ghana provides a comprehensive overview (Ntiamoa-Baidu, 1998) and is the primary source of the following account, although additional literature is also drawn on where noted. Ntiamoa Baidu's survey, conducted in 1997, included interviews with 124 bushmeat hunters and 1,262 consumers from rural and urban communities (10 villages, four towns and one city) across the forest, savanna and forest–savanna transition zones. The survey confirmed that bushmeat is a valuable natural resource in Ghana. Every year, 338,000–385,000 tonnes fresh weight is harvested, with a sales value of US$205 million to US$234 million (once dressed and smoked).[1]

One of the most important aspects of the Ghana bushmeat trade is its degree of commercialization. Overall, between 72% and 87% of all bushmeat is traded,[2] although these figures show a great deal of local variability (e.g. only 45–60% of the hunter's bag is typically sold around Ankasa and Bia protected areas; Holbech, 1998). Recent work by Brashares et al. (personal communication) has found that the proportion of bushmeat sold can vary by as much as 0–80% between villages, depending on village size and distance to paved roads and urban centres (bigger villages at closer proximities sell more). Brashares et al.'s work also highlights the importance of the local rural trade, which makes up about half of all such village sales. The remaining sales are to urban markets. Surprisingly, the national survey suggests that bushmeat consumption in rural and urban areas is quite similar. In Ghana, daily per capita rural consumption is reported at 0.033 kg per capita per day, in comparison with 0.046 kg among urban consumers. The similarity between these figures is in marked contrast to the pattern described in Central Africa, where bushmeat consumption is higher in rural areas (0.13 kg per capita per day) than in urban areas (0.013 kg; Wilkie and Carpenter, 1999). While the Ghanaian figures also vary between localities and households (e.g. rural consumption around Bia and Ankasa is 0.18–0.19 kg per capita per day among hunters' families; Holbech, 1998), they suggest that, across the country, the urban population eats about the same amount of bushmeat as the rural population (urban dwellers eat 52% of all bushmeat harvested).[3]

Bushmeat was also found to be a highly popular food in Ghana: the most popular source of animal protein in five localities (three villages and two towns) and the second most popular (after chicken, and just ahead of fish) when preferences were collated across all 15 localities. However, the price of bushmeat meant that it was eaten only infrequently. Rather, the main source of animal protein was usually fish (68% of respondents), followed by beef (15%) and chicken (6%). Bushmeat was the main source of animal protein in only 5% of cases (followed by goat (4%) and a range of other sources (2%)). Curiously, these patterns do not match well with national statistics on bushmeat, fish and beef production, estimated at 362,000 tonnes (the mid-range of the two figures cited above), 491,000 tonnes and 33,000 tonnes respectively (figures from

Ntiamoa-Baidu, 1998). It is clearly anomalous that fish and beef should be reported as the main source of protein 10 times and three times, respectively, more often than bushmeat when fish and beef production in Ghana are reported to be only one-third more than, or one-tenth of, bushmeat production. One possible explanation is that the bushmeat production figures include informal-economy production, but that the corresponding figures for fish and beef do not. However, further research into this area is required.

The bushmeat trade in Ghana is largely unregulated by either state or local institutions (*see also* Holbech, 1998). With respect to state regulation, bushmeat utilization is governed by the Wild Animal Preservation Act 1961 (Act 43), which specifies those species that are protected from hunting and the extent of this protection by age class (young or adult) and by time period (all year or only in the closed season, from August to November). However, local knowledge of the wildlife legislation, such as the timing of the closed season, is extremely poor. Lack of knowledge and compliance in bushmeat hunting and trade is attributed to the extremely limited infrastructure of the Ghana Wildlife Department, where budgetary allocation is low and most professional posts are unfilled (*see also* Tutu *et al*, 1996). In the case of local institutions, traditional authorities play a minimal role. Similarly, hunters' unions, guilds and associations, and corresponding organizations for urban bushmeat traders (e.g. the Kumasi Bushmeat Sellers' Association; Addo *et al.*, 1994) are scarce in Ghana.

The massive scale of the Ghana bushmeat trade has had a severe impact on wildlife populations. Over the past four decades, serious concern has repeatedly been raised about the sustainability of the national bushmeat harvest (e.g. Asibey, 1966; 1974; Manu, 1987; Martin, 1991). In the national survey, over 80% of all bushmeat hunters, traders and village elders interviewed perceived that bushmeat was becoming increasingly scarce (Ntiamoa-Baidu, 1998). This reported decline corresponds with the high rates of local extinction reported amongst mammals in Ghana's protected areas (Brashares *et al.*, 2001) and concern over the potential global extinction of a critically endangered primate, Miss Waldron's red colobus monkey (*Procolobus badius waldroni*) (Oates *et al.*, 2000; McGraw, 2005).

Linkages between bushmeat and fisheries

Our understanding of the Ghana bushmeat trade has recently been extended by a new study that has identified links between the national production of bushmeat and marine and freshwater fish. Brashares *et al.* (2004) found that, over a 28-year period (1970–98), those years characterized by a low fish supply were also associated with a substantial decline in the biomass of 41 mammalian species across six Ghanaian savanna protected areas. An increase in bushmeat demand, in response to a scarcity of fish, appeared to drive this pattern.

Alternative factors, such as climate, political cycles, oil prices and gross national product, did not appear to be related to these patterns. Moreover, years of low fish production were also associated with an increase in the number of hunters detected in the protected areas, and the correlation between fish supply and mammal biomass was strongest where local reliance on fish production was greatest, i.e. in those protected areas closest to the coast. Further investigation at a local scale verified that bushmeat and fish availability in local markets were linked. Specifically, patterns of bushmeat and fish sales in 12 village markets in Ghana monitored between 1999 and 2003 demonstrated that more bushmeat was sold when fish was scarce or more expensive.

Taken together, these findings suggest that bushmeat and fish production in Ghana are linked. Moreover, these findings are consistent with the results of the national survey in that bushmeat and fish were ranked similarly according to consumer preferences, and were apparently harvested in similar volumes. However, the national survey also reported that fish was the main source of animal protein for consumers 10 times more often than bushmeat (see above), a pattern that is less consistent with the preceding observations. The resolution of these inconsistencies is likely to require further study. Nevertheless, at present, the weight of evidence is clearly in favour of an important linkage between bushmeat and fisheries production in Ghana.

In light of this linkage and the importance of bushmeat and fish resources in Ghana, it is alarming to consider that in recent years mammalian biomass declined by 76% in the six protected areas monitored in this study (1970–98), while marine fish biomass in the Gulf of Guinea is also estimated to have declined by at least 50% (1977–90). At the same time, human populations have increased threefold in the region since 1970. The combination of a dwindling resource base and growing human population is a potentially catastrophic one, and improved protection and regulation of the use of these resources, together with improved livestock production, is likely to be essential to prevent further declines in the coming years. Further discussion and commentary on Brashares *et al.*'s (2004) study is provided by Watson and Brashares (2004) and Rowcliffe *et al.* (2005).

The Takoradi bushmeat commodity chain

The urban trade is a strong driver of bushmeat exploitation in Ghana (see above), but relatively little is known about how bushmeat reaches the urban markets, the livelihoods of the actors involved, and the current pattern of sustainability of this trade.[4] In light of this, our purpose here is to provide a detailed account of the bushmeat commodity chain that supplies Takoradi, a major city in Ghana. This case study is drawn from a field study in 2000 reported in a series of recent papers (Mendelson *et al.*, 2003; Cowlishaw *et al.*, 2005a,b), and readers are referred to those publications for further information.

Takoradi is the capital of Ghana's Western Region, and the country's third largest city. The present landscape around Takoradi, once tropical forest, has been shaped by a long history of human land use. This has created a complex landscape of plantations, mixed bush fallow (mainly cocoa, coconut and oil palm) and degraded tropical forest. Takoradi itself has grown rapidly over the last 100 years through the development of gold mining, railways and harbour facilities. Nevertheless, it is a relatively poor urban community where low pay and manual labour are typical features of local life (Jeffries, 1978). Bushmeat is hunted locally and sold in the markets and chop bars (cafés) of Takoradi. Monthly retail bushmeat sales in the city are 16,000 kg fresh weight (retail value US$48,000), although a much larger volume of bushmeat is captured and sold across the Takoradi catchment as a whole. Once rural sales, informal urban sales and hunters' household consumption are accounted for, the total annual harvest in the catchment for the 10 main species of mammal in the urban trade (see below) is estimated at about 1,130 tonnes fresh weight.

Ten terrestrial mammals make up 84% of Takoradi bushmeat sales by weight, with two species, the cane rat (grasscutter) and brushtailed porcupine, together contributing 50% (Table 2.1). Most bushmeat is sold processed, i.e. dressed and smoked (85% of retail sales), which extends its shelf life. Smoked meat is more expensive than fresh meat, a common pattern in Ghana's urban (Falconer, 1992) and rural areas (Holbech, 1998), presumably because smoking bushmeat requires time and money (to pay for charcoal), and is usually associated with preparation for long journeys (and so higher transport costs). It is also possible that consumers prefer the taste of smoked meat. Bushmeat is eaten throughout the year, particularly during festivals and holidays, but it tends to be purchased only in small amounts because of its expense. The market price of smoked bushmeat, which is correlated with consumer taste preferences, the distance the carcass has travelled to market and carcass size,[5] is highest for cane rat (US$2.97/kg). In contrast, domestic meat and fish are less expensive and more widely eaten. The cheapest alternative to bushmeat is fish (US$1.34/kg), followed by chicken (US$1.68/kg), pork (US$1.74/kg), beef (US$1.88/kg) and mutton (US$2.15/kg). The high price of bushmeat is typical of urban markets in Ghana (Falconer, 1992; Asibey, 1966) and across the region (Martin, 1983; Steel, 1994). This pattern is at least partly likely to reflect its limited supply: bushmeat is produced at low volume and at considerable distances from urban centres.

Actors and trade

There are five actors in the Takoradi commodity chain: the producers (commercial hunters and farmer hunters), wholesalers and retailers (market traders and chop bar owners). All hunters are men, whereas all urban traders are women, a common pattern in the bushmeat trade in Ghana (e.g. Falconer,

Table 2.1 **Species traded in the Takoradi bushmeat commodity chain.**

Taxon	Proportion of fresh biomass	Market price* (US$/kg)	Distance travelled* (km)
Rodents			
Cane rat (*Thryonomys swinderianus*)	0.31	2.97	48
Brush-tailed porcupine (*Atherurus africanus*)	0.19	2.27	34
Giant rat (*Cricetomys emini*)	0.06	2.35	36
Ground squirrels, multiple spp.	0.02	1.90	54
Flying squirrels (*Anomalurus* spp.)	0.01	2.59	112
Ungulates			
Bushbuck (*Tragelaphus scriptus*)	0.09	2.63	65
Bay duiker (*Cephalophus dorsalis*)	0.05	2.59	82
Black duiker (*Cephalophus niger*)	0.05	2.55	104
Maxwell's duiker (*Cephalophus maxwelli*)	0.03	2.59	96
Royal antelope (*Neotragus pygmaeus*)	0.03	2.67	105
Others			
Giant snail (*Achatina marginata*)	0.06	2.10[†]	45
Snails, multiple spp.	0.06	1.94[†]	44
Bush crabs, multiple spp.	0.03	1.72[†]	23
Crested guinea fowl (*Guttera edouardi*)	0.01	1.75	41
Fruit bat (*Eidolon helvum*)	<0.01	1.53	35
Green fruit pigeon (*Treron calva*)	<0.01	1.65	133

*Mean values.
[†]Prices are given for fresh meat because invertebrates were never smoked before sale.

1992; Ntiamoa-Baidu, 1998), the bushmeat trade in West Africa (Liberia: Anstey, 1991; Ivory Coast: Caspary, 1999) and the wider West African trade in agricultural produce (e.g. Robertson, 1984; Guyer, 1987).

Commercial (full-time) hunters depend on bushmeat as their primary source of income, whereas farmer (part-time) hunters capture bushmeat to supplement their income from agricultural produce and to control crop pests. Most Takoradi hunters use wire snares rather than shotguns to capture their prey, although farmer hunters also sell invertebrates gathered by women and children in the household (e.g. Falconer, 1992). The hunters sell their catch to urban traders, primarily wholesalers, but also to market traders and chop bars (Figure 2.1). Wholesalers are bulk traders who work from home (there is no dedicated wholesale bushmeat market in Takoradi). Wholesalers aim to sell their bushmeat within 4 days of purchase from the supplier, but will also freeze their surplus when appropriate. Wholesalers sell to market traders

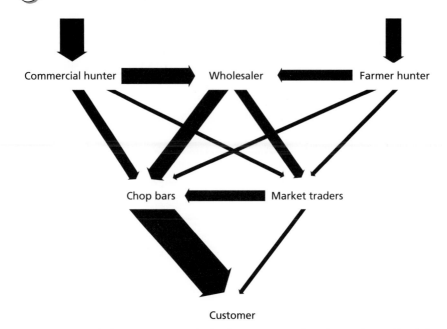

Figure 2.1   **Trade flow patterns in the bushmeat commodity chain. The width of the arrows is proportional to the volume of fresh biomass traded. The total biomass traded in this figure is 16,000 kg across all species over a 4-week period. From Cowlishaw *et al.* (2005b).**

and chop bars (Figure 2.1), and each individually handles about 4% of the entire bushmeat trade (Table 2.2). Market traders sell from stalls in Takoradi central market. Some are part-time, and also work as cleaners, shop assistants or seamstresses. However, all hope to trade full-time in the future. Although market traders as a group handle 41% of the trade, they sell only 15% to the public because most of their sales are to chop bars (Figure 2.1 and Table 2.2). Chop bars (cafés) are usually situated close to large workplaces such as factories, council buildings and lorry stations, and vary in size from a small room for about 10 people to a large building that can seat over 100. Owners employ a variety of staff: a medium-sized chop bar (seating between 20 and 50 people) might employ a general assistant, several cooks, waitresses, one or two cleaners and at least one fufu pounder. All of these staff (except fufu pounders) would be women. Chop bar operators are the most numerous actor group and sell most bushmeat to the public (Figure 2.1 and Table 2.2).

The primary route of bushmeat is from commercial hunters to chop bars via wholesalers, but there is substantial trade along alternative routes (Figure 2.1) and considerable freedom in the network: all actor groups trade with one another, and individuals are free to trade with whomever they wish. Hunters usually obtain their best price from market traders, so the central market is their first stop. Here they will move between stalls until their price has been

Table 2.2 **Actors in the Takoradi bushmeat commodity chain.**[*]

Actor group	Age (years)	Education (proportion attending school)[†]	Years in trade	Number of actors estimated in trade	Fraction of trade handled (by weight)		Fraction of sales incurring transport costs[‡]
					Per group	Per actor	
Farmer hunter	32	0.33	17	75	0.38	0.005	0.59
Commercial hunter	25	0.67	8	50	0.62	0.012	0.78
Wholesaler	49	0.60	18	14	0.58	0.041	0.49
Market trader	31	0.55	10	16	0.41	0.026	0.19
Chop bar owners	36	0.96	17	143	0.85	0.006	0.00

[*]The sample sizes for respondents are 3, 9, 5, 11 and 24 for farmer hunters, commercial hunters, wholesalers, market traders, and chop bar owners, respectively. All figures in years are median values across the sample.

[†]For those attending school, the median age at completion was 9 years (farmer hunters, market traders) or 10 years (all others).

[‡]Most transport costs for hunters relate to transport into town, whereas most costs for wholesalers and market traders refer to transport costs around town (to make deliveries to retailers).

met, or else leave the market to sell elsewhere. In contrast, hunters obtain their lowest price from wholesalers, but are usually able to sell their entire bag in such transactions. No actor group appears to exert power over other groups in the trade, and there are no associations or guilds. Even the position of 'market queen', a trader elected by other traders in a commodity group to act as their general representative and an arbitrator in disputes (Clark, 1994), which is common in other agricultural markets in Ghana (e.g. Hinderink and Sterkenberg, 1975; Lyon, 1999), is lacking for bushmeat traders in Takoradi.

Livelihoods and income

To estimate the gross profits acquired through the bushmeat trade, we compared the sales price per kilogram of six species that together constitute more than half (53%) of the trade[6] for each of the actor groups. This comparison indicates that the proportion of the final chop bar sales price captured by each actor was greatest for hunters (73%), followed by the chop bars themselves (18%), market traders (5%) and wholesalers (3%). These findings are consistent with those of previous studies in Ghana. In Sunyani city, for example, hunters obtain 67% of the chop bar sales price (Asibey, cited in Martin, 1991). Between 1977 and 1978, it was estimated that hunters earned the same income as a civil servant (Asibey, 1977) or 8.6 times the average salary of a government labourer (Martin, cited in Holbech, 1998), whilst 20 years later Ntiamoa-Baidu (1998) found that hunter income was the same as an entry-level graduate wildlife officer and 3.5 times the government minimum wage.

The Takoradi figures for gross profit do not take into account the operating costs of participation in the bushmeat trade. A qualitative assessment of these expenses indicates considerable variation between actor groups (Table 2.3). Most noticeably, however, these costs tend to be lowest amongst hunters, as capital investment is low (snares are cheap), and neither employees nor premises are required. The most important expense for hunters is the cost of transportation of their bag into market (although these expenses are sometimes covered by wholesalers; Table 2.2). This involves a journey by taxi, bus or lorry (there are no local specialist transporters), with costs proportional to the distance travelled and the sales value of the load.[7] If we incorporate the costs of transport into the figures above, the fraction of the final sales prices acquired by hunters is reduced, but the general pattern remains the same and hunters still capture most of the final sales price: hunters (65%), chop bars (18%), market traders (5%) and wholesalers (3%). Urban actors, in contrast, appear to experience much more significant operation costs than hunters. In addition to the expense of purchasing bushmeat from suppliers, these consist of capital assets for storage (fridges and freezers) and smoking (ovens and grills), and the running costs associated with business premises and employees (Table 2.3).

Table 2.3   **Financial costs of participation in the Takoradi bushmeat commodity chain.**

Expense	Relative expense			
	Hunters	**Wholesalers**	**Market traders**	**Chop bar owners**
Equipment	Low	High	Moderate	High
Transport	High	Moderate	Low	None
Employees	None	Low	Low	High
Premises	None	None	Low	Moderate
Overall	*Very low*	*Moderate*	*Low*	*High*

Actors can, to some extent, minimize their operating costs through the support of their extended family (*see also* Clark, 1994). Amongst farmer hunters, for example, relatives can establish informal cooperatives that minimize the costs of both harvesting (e.g. by sharing a firearm) and transport (e.g. by nominating one member to carry the collective bag to market). However, hunter costs are already relatively low, and the role of kin support networks appears to be much more significant for urban traders. At the outset, the start-up costs of city businesses are often mitigated through inheritance: 40–45% of Takoradi wholesalers, market traders and chop bar owners inherit their businesses from female relatives, usually the mother. Further assistance is also provided in a variety of forms, including transport and financial credit, but especially labour: mothers, sisters, daughters and other female relatives often cover for one another on market stalls, while the relatives of chop bar owners can be found in almost every staff position. This support is likely to make an important difference to the viability of urban businesses, although this assistance is likely to only partly offset the costs of trading described above.

The livelihood aspirations of the different actors groups are consistent with these patterns of income. In general, urban actors consider the bushmeat trade a difficult profession of relatively low status, and one that does not offer a secure future. Apart from those wholesalers and chop bar owners who have been very successful, most traders hope that their daughters will pursue more respectable professions. In the case of hunters, many commercial hunters hope to become farmers (and so presumably farmer hunters). Tenure on farmland provides more control over hunting, and, most importantly, the farm itself acts as a secure and significant source of food and income.

Biological sustainability

Takoradi is supplied with bushmeat from a large catchment around the city: on average, all species are harvested at distances greater than 20 km from the city, and eight species come from an area more than 50 km away (Table 2.1).

These animals are hunted in a predominantly agricultural mosaic landscape, incorporating fallow fields and remnant forest areas: a 'farmbush' matrix.

The biological sustainability of the Takoradi bushmeat trade is of fundamental importance. Unfortunately, the direct assessment of sustainability in such systems is very difficult because of the detailed data required on the production and off-take of prey species across large spatial and temporal scales. To circumvent this problem, indirect indicators can be used. Four lines of indirect evidence collectively suggest that the Takoradi trade is currently sustainable. First, the annual harvest of terrestrial mammals does not exceed the estimated sustainable yield in the catchment. Second, larger mammals are not more heavily depleted closer to the city than they are away from it. Third, the price of bushmeat has not outstripped inflation over the last four decades. Fourth, the prices of alternatives, such as domestic meat and fish, have not fallen relative to the price of bushmeat (over the same period).

Although this might seem to be good news, the current pattern of sustainability appears to be the outcome of historical overexploitation. A comparison of local species that are present and absent in the Takoradi market indicates that all those with low reproductive rates (namely primates and large ungulates), and hence a poor ability to sustain high levels of hunting, are absent from the trade (Table 2.4).[8] The absence of these species suggests that they have already been harvested to low levels and may now be locally extinct. Following the disappearance of these taxa, the only species that are now left to supply the market are species with high reproductive rates (primarily rodents and small ungulates) that can sustain heavy exploitation. This explanation of a historical decline is also supported by interviews with hunters: only hunters that have been active for 8 years or more perceived a decline in prey abundance; those hunters active over shorter time periods did not report any changes. The Takoradi bushmeat trade might therefore be characterized as one of 'post-depletion sustainability'.

This conclusion is an important one and merits two qualifications. First, these circumstances may not persist indefinitely. Changes in current conditions, such as the habitat quality of the catchment or the size of the urban population, are likely to influence the supply and demand of bushmeat and so affect the pattern of sustainability. Second, the extent to which these findings can be generalized to other localities is also uncertain. In particular, Takoradi's coastal location makes fish a widely available substitute, which might reduce bushmeat consumption. However, bushmeat and fish are unlikely to be interchangeable in Takoradi because in this locality the former is a luxury good whereas the latter is not. Similarly, although Takoradi has a relatively low per capita bushmeat intake for Ghana (0.01 kg per day), this figure is typical for urban dwellers across Central Africa (see above), indicating that Takoradi may be a representative model in this respect. Nevertheless, further research would clearly be helpful in clarifying these areas of uncertainty. In any case,

Table 2.4 **Takoradi market presence, and the population growth rates (r_{max}), of a selection of primate, ungulate and rodent species.**

Species	Mass (kg)	r_{max}	Presence in market
Flying squirrels (*Anomalurus* spp.)	0.7	1.09	Present
Pel's flying squirrel (*Anomalurus pelii*)	1.5	0.85	Present
Royal antelope (*Neotragus pygmaeus*)	3	0.68	Present
Brush-tailed porcupine (*Atherurus africanus*)	3	0.68	Present
Maxwell's duiker (*Cephalophus maxwelli*)	10	0.46	Present
Black duiker (*Cephalophus niger*)	20	0.37	Present
Bay duiker (*Cephalophus dorsalis*)	20	0.37	Present
Bushbuck (*Tragelaphus scriptus*)	42	0.29	Present
Yellow-backed duiker (*Cephalophus sylvicultor*)	63	0.26	Absent
Red river hog (*Potamochoerus porcus*)	80	0.24	Absent
Giant hog (*Hylochoerus meinertzhagene*)	150	0.20	Absent
Lesser spot-nosed monkey (*Cercopithecus petaurista*)	4	0.18	Absent
Campbell's monkey (*Cercopithecus campbelli*)	5	0.17	Absent
African buffalo (*Syncerus caffer*)	285	0.16	Absent
Red-capped mangabey (*Cercocebus torquatus*)	7	0.15	Absent

similarities between the bushmeat trade in Takoradi and other localities in Ghana suggest that the historical depletion of vulnerable species, followed by a trade in more robust species, may be a typical pattern. The same five bushmeat species that make up 67% of the Takoradi market biomass (cane rat, brush-tailed porcupine, Maxwell's duiker, bushbuck and black duiker) also constitute 70% of market biomass across 15 localities in five different regions (Ntiamoa-Baidu, 1998).

Concluding comments

The bushmeat trade is a well-established multimillion-dollar industry in Ghana that is important for both rural and urban livelihoods. However, the use of wild animal populations to fuel this industry has not been sustainable. This has led to widespread local extinctions, particularly of more charismatic species such as primates and large ungulates. These extinctions not only threaten the future of the bushmeat trade, but will have wider long-term impacts, including the loss of ecosystem services (e.g. seed dispersal for forests) and the loss of tourism revenues from wildlife viewing. The human and environmental costs of these impacts still remain to be calculated. Moreover, the problem is exacerbated by interactions with another food production system, namely fisheries, which are also in decline. There is therefore an urgent need for effective action to make the bushmeat trade in Ghana sustainable, to ensure the long-term availability of this renewable resource to the people of Ghana and to protect those species that are threatened by it.

The findings of our Takoradi case study indicate one possible way forward. Our results suggest that bushmeat markets can be sustainably supplied solely from robust species, and that these can be drawn from an agricultural mosaic landscape (a farmbush matrix). This indicates that, at least in principle, it is possible to support the livelihoods of those people who depend on the bushmeat trade while also protecting threatened species and habitats. Most previous assessments of the sustainability of the bushmeat trade in other parts of the world have focused on tropical forest habitats, which are characterized by low animal productivity (e.g. Robinson and Bennett, 2000). Agricultural landscapes are likely to be much more productive, due to the presence of crops and patches of vigorous secondary forest growth (e.g. Robinson and Bennett, 2004). In combination with the need to control crop pests, this productivity is likely to explain why bushmeat extraction in secondary forest can match or exceed that in primary forest (Wilkie, 1989) and why Ghanaian hunters prefer to hunt in farmbush matrix habitats (Falconer, 1992; Holbech, 1998).

However, the implementation of systems that allow an active bushmeat trade alongside the coexistence of vulnerable species will be difficult in practice because of limited institutional capacity. In Ghana, wildlife laws already exist to regulate the bushmeat trade, many of which are consistent with the approach outlined here, but public awareness and state enforcement of these laws is extremely limited (a common pattern in the region; cf. Democratic Republic of Congo; Rowcliffe *et al.*, 2004). Wider issues of capacity building and good governance will therefore also need to be tackled if the bushmeat trade is to be managed successfully (Davies, 2002).

From a livelihoods perspective, there are perhaps three important conclusions to draw. First, the income that hunters derive from bushmeat sales indicates that bushmeat has the potential to contribute to poverty alleviation at the household level. A similar finding has also been reported in Democratic Republic of Congo (de Merode *et al.*, 2004). Further analysis of how Ghanaian farmers might balance their agricultural and hunting activities, and the associated impacts on prey populations, is presented by Damania *et al.* (2005). Second, individual bushmeat businesses show many similarities to those of agricultural commodities in West Africa, in terms of both scale and pattern of operation (see Fafchamps and Gabre-Madhin, 2001). Interventions aimed at improving the livelihoods of those dependent on bushmeat might therefore be derived from those used for agricultural traders in the region, although with the crucial proviso that bushmeat can never be produced in the same volume as most agricultural commodities, even when harvested sustainably. The third and final point to consider is that the livelihoods of different actor groups in the commodity chain are tightly interwoven with those of others, and interventions at one point in the chain are likely to have consequences elsewhere. Schemes to improve livelihoods or manage the trade for sustainability should therefore engage with all actor groups.

Acknowledgements

We thank Lars Holbech, Katherine Homewood, Francis Hurst, Catherine MacKenzie, Candy Mends and Paul Symonds for their assistance during the Takoradi study; Tom Brass, David Brown, Glyn Davies, Lars Holbech, E. J. Milner-Gulland, John Robinson, David Wilkie and several anonymous referees for their valuable comments on earlier papers; Justin Brashares for his comments on this book chapter and permission to cite some of his unpublished research findings, and Glyn Davies and David Brown for their invitation to speak at the ZSL/ODI Bushmeat and Livelihoods conference and to contribute to this book. We also extend a special thanks to the many actors in the Takoradi bushmeat commodity chain who participated in this study. The project was funded by the UK Natural Environment Research Council (NERC) and Economic and Social Research Council. G.C. is currently in receipt of an NERC Advanced Fellowship. The field work was carried out in affiliation with the Protected Area Development Programme of the Wildlife Department, Ministry of Lands and Forestry, Republic of Ghana. This chapter is a contribution to the ZSL Institute of Zoology Bushmeat Research Programme.

Notes

1 These figures for bushmeat volume are derived from estimates of the national harvest (385,000 tonnes) and consumption (225,000 tonnes) based on interviews with 98 active hunters and 1,262 consumers respectively. As harvest volumes are fresh weights, but consumers eat meat only after it has been dressed and smoked (i.e. processed), the consumption volume has been converted to fresh weight (i.e. 338,000 tonnes) (using a conversion factor of 1.5, as carcasses that have been dressed and smoked weigh about two-thirds of the fresh weight). A third volume estimate is also available for the amount of processed meat sold by bushmeat traders (92,000 tonnes), but this is based on a very small sample of traders and therefore not considered further here. Financial values are estimated on the basis of processed weights, as this is the state in which most bushmeat is sold. The differences between the hunters' and consumers' US$ figures may at least partly reflect the fact that hunters do not sell all their bag, but rather eat some of it in their own household and give some of it away as gifts.

2 The first figure derives from purchases by consumers ($n = 1,262$ respondents), whereas the second comes from sales by active hunters, i.e. hunters who had been hunting within the month preceding the survey ($n = 98$ respondents).

3 This figure is based on the per capita consumption in rural and urban areas, according to the size of the rural and urban population. Data for urbanization in Ghana are taken from the Population Division of the Department of Economic and Social

Affairs of the United Nations Secretariat, 2004, World Urbanization Prospects: The 2003 Revision, Urban and Rural Areas Dataset (POP/DB/WUP/Rev.2003/Table A.7), New York: United Nations (dataset in digital form available online at www.un.org/esa/population/ordering.htm). Data for the national population of Ghana are taken from the Population Division of the Department of Economic and Social Affairs of the United Nations Secretariat, 2003, World Population Prospects: the 2002 Revision, dataset on CD-ROM, New York: United Nations (available online at www.un.org/esa/population/ordering.htm).

4 Although some work has been carried out on the bushmeat trade in Kumasi, Ghana's second largest city, this market is renowned for its size (it has been described as 'perhaps the largest in West Africa'; Clark 1994, p. 1). Consequently, it may not be a typical model for other urban markets. Previous studies on the Kumasi bushmeat trade include Addo et al. (1994) and Falconer (1992).

5 We investigated three potential determinants of market price: carcass size, consumer taste preferences and travel costs associated with the capture distance from market. All three variables were correlated with market price across taxa: carcass size (Pearson correlation $r_p = 0.71$, $n = 14$, $P = 0.005$), taste preferences ($r_p = 0.93$, $n = 8$, $P = 0.001$) and journey distance ($r_p = 0.64$, $n = 14$, $P = 0.013$). Carcass size and journey distance were also weakly correlated with one another ($r_p = 0.53$, $n = 14$, $P = 0.05$), which makes their independent effects difficult to disentangle with this small sample, but the effect of taste preference was independent of both size and distance ($r_p < 0.50$, $n = 8$, $P > 0.20$ in both cases). Note that this result slightly differs from, and supersedes, that presented in Cowlishaw et al. (2005b).

6 These six species are cane rat, giant rat, bay duiker, black duiker, Maxwell's duiker and royal antelope. The comparison was made for smoked meat.

7 Distance travelled, D (km), and the total sales value, V (US\$), had significant positive effects on transport costs (stepwise regression: step 1, D enters model: cumulative $r^2 = 0.62$, $P < 0.001$; step 2, V enters model: cumulative $r^2 = 0.73$, $P = 0.009$; whole model: $F_{2,17} = 27.0$, $P < 0.001$). Other variables, namely the mass of the load (which correlated with V: $r_p = 0.94$, $n = 30$, $P < 0.001$) and the number of deliveries made on the journey, did not have a significant effect.

8 To control for the potentially confounding effects of species distribution and habitat preferences, all species have a local distribution and occur in farmbush matrix habitats (Grubb et al., 1998). Similarly, to control for the legality of trade, all species are on the Second Schedule of the Wildlife Laws of Ghana (Government of Ghana, 1998), and so can be legally hunted (as adults unaccompanied by young) during the study period (i.e. outside the closed season). All species of weight ≥ 700 g are considered, where species body mass is taken from Kingdon (1997), and the intrinsic rate of population increase, R_{max}, is calculated following Rowcliffe et al. (2003). Other members of these three taxonomic groups that were recorded in the Takoradi market (Table 2.1) that met the ≥ 700 g size criterion are not listed here, either because they occur on a different schedule (e.g. giant rat, Third Schedule), or because they can be hunted without restriction (the cane rat). Note that C. campbelli is classified with C. mona in the Wildlife Laws of Ghana.

$$\textbf{3}$$

Bushmeat Markets – White Elephants or Red Herrings?

John E. Fa

Introduction

Bushmeat, the meat of wild animals, is a highly valuable non-timber forest product in West and Central Africa. Estimated extraction rates for a large number of wildlife species in these regions are considered well above production levels (Fa *et al.*, 2002a). Intensity of hunting of wild species has risen because of the transition from subsistence to commercial (or market) hunting and trading of wildlife, this conversion having occurred some time ago; for example, Hart (1978; 2000) observed it among the Mbuti of the Democratic Republic of Congo (DRC) in the 1950s. Market hunting is distinguished from subsistence hunting because it is not restricted to supplying the hunter's family with meat; instead hunters sell products to the open market. However, as indigenous groups supplement their incomes with the sale of meat, the formerly benign levels of subsistence hunting are now interlaced with essentially commercial activities that are having a negative impact on biodiversity. The bushmeat trade applies to all wildlife from small to large animals inhabiting natural habitats that are shot or trapped and sold for public consumption either fresh or preserved, usually by smoking. Wildlife trade across national borders is generally small.

The bushmeat trade is currently of great interest to development and conservation agencies, due to concerns over the sustainability of its use and the implications of its loss for poor rural households (Fa and Peres, 2001; Brown and Williams, 2003). The general assumption is that bushmeat trade has increased as a result of rising human populations and the collapse of economic infrastructure, but also the need for animal protein (Fa *et al.*, 2003). Greater accessibility to formerly impenetrable forest areas, as well as improving hunting technologies and the lack of law enforcement, has further intensified the trade (Robinson and Bennett, 2000). The magnitude of the trade has been valued between US$42 million US$205 million across countries in West and Central Africa (Davies, 2002). In Gabon, Feer (1993) estimated that around 20% of game consumed annually was from commercial hunting, and in a more recent study > 13,000 tonnes of wild meat was reportedly sold in urban markets (Wilkie *et al.*, 2005).

Published hunting studies are useful for determining trends, but, because they generally cover small geographical areas, they are of limited use for understanding wildlife extraction within entire regions. Although some data are available on harvest rates and population status of hunted prey species (Fa et al., 2005), there is little information on region-wide impacts. This is because estimating and monitoring changes in abundance of forest animals is costly and often too time and energy intensive to apply at larger spatial scales. Information is usually available to assess how sustainable the trade is in a particular area. Population estimates of bushmeat species are difficult and time-consuming (and therefore expensive) to obtain, particularly in hunted areas where densities are low and animals are wary of humans (Fitzgibbon, 1998). Data on hunter behaviour and off-take rates are confined to sites where researchers have invested substantial time and effort (for example, Muchaal and Ngandjui, 1999).

Juste et al. (1995), Fa et al. (2000) and later Rowcliffe et al. (2003) have suggested that because bushmeat markets are found in almost every town and village in a region, and are important concentration centres of wildlife harvests, they may be used as indicators of the state of hunted faunal assemblages. Although accuracy may be confounded by a number of difficult to measure variables (see below), market data can be useful at least as minimum extraction rates of bushmeat species. Some authors (Fa et al., 2000; Rowcliffe et al., 2003; Crookes et al., 2006) propose that market data can be used to indicate faunal depletion, by inferring the level of overexploitation of bushmeat in the supply areas from the proportion of particular species in the market. This is based on the observation that an 'extinction filter' usually operates (Balmford, 1996). As environments become more depleted, smaller, more productive, species would replace larger-bodied species of low productivity. This phenomenon of hunting down the size classes is also well known in fisheries (C. M. Roberts, 1997). Thus, local extinction events in prey species assemblages is a non-random process governed by morphological, metabolic and reproductive traits, which are usually correlated with body mass (R. H. Peters, 1983). Large body size is one of the most frequently cited traits facilitating extinction events, probably because it is associated with low reproductive rates, low population densities, long generation time and long lifespans (Hennemann, 1983). In Amazonia, Jerozolimski and Peres (2003) detected a gradual shift from large- to small-bodied prey species.

In this chapter, I argue that this basic observation, which has been used in the past to assess faunal depletion scenarios from market data (see Fa et al., 2004; Cowlishaw et al., 2005b), can be employed to assess hunting impact at both temporal and spatial scales, and moreover within relatively large geographical areas. From available published and unpublished data, I first review the basic characteristics of bushmeat markets in West and Central Africa in terms not just of species composition, but also of volume traded. Much of the focus is on mammals because this class of vertebrates constitutes the largest proportion of hunting kills in African moist forests. Finally, I ask the question whether realistically, given the often incomplete knowledge of variables surrounding bushmeat markets, the data that emerges from these can be played up as indicators of the state of the surrounding wildlife populations, effectively becoming a *white elephant*.[1] In contrast, I also discuss whether monitoring bushmeat markets could be in fact irrelevant, or a *red herring*[2] to resolving the bushmeat trade issue.

General characteristics of the bushmeat trade and markets

Bushmeat is consumed for many reasons (Schenck *et al.*, 2006). Cultural preference and availability allow bushmeat to supply food and economic revenue. In most situations, it is men who hunt but women who take charge of all the downstream processing and commerce, to the point of sale. Basically, wild meat is sold as fresh carcasses or smoked meat in markets, at roadsides, in hunters' homes or as cooked dishes in chop bars (cafés). Bushmeat is traded at a number of entry points in the commercial chain, from the hunter to the consumer, and hunters may sell their kill as whole animals to a trader or a chop bar operator, who then retails it in smaller pieces. In some cases, hunters may dress the carcass and sell pieces direct to consumers in the village. Hunters or their emissaries may carry the meat to the point of sale, but in the case of professional hunters operating from hunting camps traders may travel to the camp to buy the smoked meat. In the cities, bushmeat may pass through middlemen before it reaches the consumer.

The main concentration sites for the sale of bushmeat, on a regular basis, are without any doubt within markets. Such public gatherings, where the buying and selling of merchandise including bushmeat take place, occur in almost every sizeable village or town. Bushmeat can be traded and displayed on make-shift counters, or in larger cities on more permanent stalls within purpose-built market buildings (Figure 3.1). Some, like the Atwemonon market in Ghana (Ntiamoa-Baidu, 1998; Crookes *et al.*, 2006), are highly organized and the bushmeat trade and associated chain of chop bars are handled as small-scale family businesses handed down from parents to children.

In all studied areas (Fa, 2000; Cowlishaw *et al.*, 2005b; East *et al.*, 2005), there are five main actor groups identified in the bushmeat trade: farmer hunters, commercial hunters, wholesalers, market traders and chop bar operators. All hunters are men, whilst all other actors are women. Hunters live and work in the rural areas and capture their prey using snares and shotguns. Commercial hunters depend entirely on bushmeat for their livelihood, whereas farmer hunters sell bushmeat to supplement their income from agricultural produce. The women traders – wholesalers, market traders and chop bar operators – live and work in the city. Wholesalers work from home. They buy meat in bulk from the hunters and sell to the retailers: the market traders and the chop

Figure 3.1 **A typical make-shift market in Nigeria.**

bars. Market traders operate from stalls in the market, whereas chop bars are scattered across the city. Women form the main clientele for market traders, whereas men are more likely to frequent chop bars. The primary route of trade is from commercial hunters to chop bars via wholesalers, although there is also substantial trade along other routes. Each trader has her own set of hunters who supply her with meat and whom she prefinances by granting loans. The trade provides income for a large number of people, hunters and traders, but it is a fairly closed system.

Most bushmeat markets are largely unregulated by either state or local institutions. In a number of countries, some wildlife species (e.g. endangered species) nominally protected from hunting by legislation are still consumed as bushmeat. Bushmeat sold openly to the public is a typical feature of many African countries, and markets are found in almost every village or town in the region. Bushmeat marketing is particularly well developed in West and Central Africa, which is also the area where the trade has been well documented since as long ago as the 1970s (see Asibey, 1977; Jeffrey, 1977).

Composition of bushmeat species traded

Although a variety of wildlife ranging from insects to vertebrates (amphibians, fish, reptiles, birds and mammals) is consumed, in terms of weight and numbers mammals make up the bulk of the bushmeat trade. In the Cross-Sanaga region in Nigeria and Cameroon, for example, Fa *et al.* (2006) calculated that, of over a million carcasses traded in 100 sites, 99% were mammals, of which around 40% were ungulates (duikers and pigs), 30% rodents and close to 15% primates. Information on bushmeat volume traded within markets in the moist forest areas has been published for Ghana (Ntiamoa-Baidu, 1998; Cowlishaw *et al.*, 2005b; Crookes *et al.*, 2006), Bioko (Fa *et al.*, 1995); Rio Muni (East *et al.*, 2005; Fa *et al.*, 1995), DRC (Colyn *et al.*, 1987), the Cross-Sanaga region (Nigeria, Cameroon) (Fa *et al.*, 2006) and Gabon (Starkey, 2004). From these sources, a ternary plot of reported numbers of mammal carcasses shows that most bushmeat markets sell largely ungulates and rodents, but primates constitute more than 20% (Figure 3.2). These three taxa are the most important for human consumption in all areas where the trade has been documented (*see also* studies in Robinson and Bennett, 2000), but significant variation in the proportions of ungulates, rodents and primates is typical. The relative contributions of these taxa are highly uneven, as often three species alone – small duikers such as *Cephalophus monticola* and *C. maxwelli*, rodents

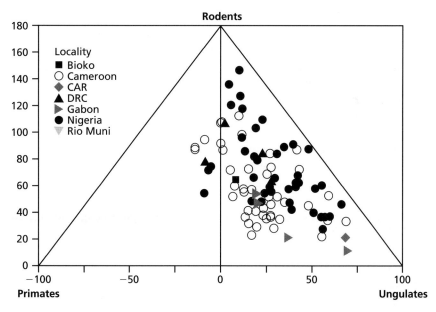

Figure 3.2 **Ternary plot or ternary graph of proportions of the three mammal taxa on sale in bushmeat markets in West and Central Africa. A ternary plot is a specialization of a barycentric plot for three variables. It graphically depicts the ratios of three proportions. It is often used in petrology, mineralogy and other physical sciences to show the relative compositions of soils and rocks, but it can be more generally applied to any system of three proportions. Data sources: Bioko, Fa *et al.* (1995); Cameroon, Fa *et al.* (2006); CAR, Noss (1995); DRC, Colyn *et al.* (1987); Gabon, Steel (1994); Nigeria, Fa *et al.* (2006); Rio Muni, Fa *et al.* (1995).**

such as the cane rat (*Thrionomys swinderianus*) and the brush tailed porcupine (*Atherurus africanus*) – constitute over 50% of the total weight traded.

Bushmeat volumes hunted and traded

Estimates of bushmeat volume taken by hunters in a variety of African moist forest sites confirm average annual figures of almost 16,000 kg per site, involving around 70 species of mammals (Fa *et al.*, 2005). Most hunting is undertaken by non-discriminatory snares, hence the emerging game profiles (body mass of animals hunted) are not biased to larger prey, as observed in equivalent habitats in South America (Fa and Peres, 2001; Jerozolimski and Peres, 2003). In a study of 42 hunters in Rio Muni (Fa and Garcia Yuste, 2001), most animals hunted (86%) were caught in snares, and only 8% were shot (more than 60% of all primates), whilst the remaining 6% were taken by other methods (e.g. machete, dogs). Ungulates, rodents and carnivores, but not reptiles or primates, were more vulnerable to snares than to firearms, 10 of the 12 ungulates, six of the seven rodents and seven of the nine carnivores being caught exclusively with snares. Thus, because of the pre-eminence of snares over guns, African hunters have their game choices made for them, and arguably what appears in hunters' bags is directly related to the relative abundance of the different prey species. A study of hunters' quarry in north-eastern Gabon (Lahm, 1993) showed that rodents and other small mammals were more likely to be consumed locally while the more appealing and profitable game were sold. Similar practices have been reported in DRC, where 76% of ungulates killed by hunters were sold whereas 55% rodents and 57% of primates were eaten in villages (Colyn *et al.*, 1987). Fa and Garcia Yuste (2001) also showed that almost 80% (range 33–100%) of the quarry in Rio Muni hunters' bags was sold to local markets, and there was a significant positive correlation between the proportion of animals sold and the species' body mass (Figure 3.3). Thus, whilst an average of 33% of the <1 to 4 kg species were sold, almost half of >4 kg animals were sold.

Snare hunting is without doubt far more severe than hunter profiles or market data indicate, as trapped animals rot or are lost to scavengers. Delvingt (1997) reported that losses accounted for between 4% and 36% of all animals trapped, and Fa and Garcia Yuste (2001) reported similar values. In another study, 26–39% of all animals were left to rot in traps in distant forests, but village traps were checked more frequently, and wastage was around 11% (Dethier, 1995).

Data on actual bushmeat volumes for sale, taken from the literature (see above), generally indicate a very large variation in amounts traded per site. From more extensive, multiple-site studies (Starkey, 2004; Wilkie *et al.*, 2005; Fa *et al.*, 2006) amounts traded ranged from about 100 to 9,000 carcasses per annum (Table 3.1). When bushmeat volume traded per site is adjusted

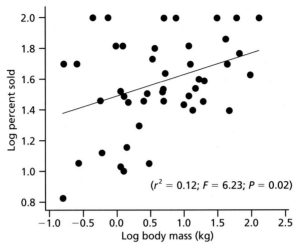

Figure 3.3 **Proportion animals sold according to the body size of species (data from Fa and Garcia Yuste, 2004).**

by the number of inhabitants in each site (data from Fa *et al.*, 2006), about 20 kg per person per annum (median 7.7, range 0.1–392.2) is available, but highly skewed, as 45% of all studied sites had between 0 and 4 kg of bushmeat per inhabitant per annum (Figure 3.4a). The more populated sites did not have more bushmeat on sale (in fact, bushmeat availability fell with larger settlements) (Figure 3.4b), but bushmeat volume on sale per site was negatively correlated with mean body mass of the animals on sale (Figure 3.4c).

Table 3.1 **Summary of main characteristics of bushmeat extracted in 89 human settlements within the Cross and Sanaga Rivers region in Nigeria and Cameroon (data from Fa *et al.*, 2006).**

Variable	Mean ± SE	Median	Minimum	Maximum
Settlement size	11,972 ± 8,241	866	55	576,306
Bushmeat biomass (kg) per annum	11,880 ± 1,660	7,820	733	76,991
Number of carcasses per annum	1,859 ± 224	1,219	128	8,822
Number of species per annum	17.5 ± 0.5	17.0	10.0	30.0
Mean body mass (kg)	6.9 ± 0.3	6.4	2.4	15.1
Distance to main national park (km)	29.5 ± 3.7	18.5	0.2	99.5
Per cent ungulates	41.7 ± 2.0	43.0	5.3	71.1
Per cent primates	14.9 ± 1.1	14.4	0.7	32.3
Per cent rodents	36.4 ± 1.9	32.5	12.4	77.9
Per cent carnivores	7.1 ± 0.8	3.9	0.0	27.4

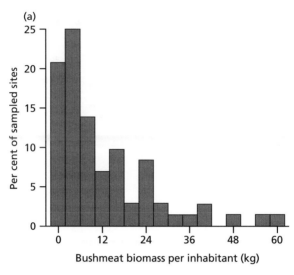

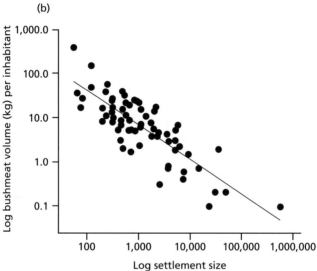

Figure 3.4　**Main characteristics of bushmeat traded in the Cross-Sanaga Rivers region: (a) frequency distribution of bushmeat volume (kg) per inhabitant available per annum within sampled sites; and log–log relationships between (b) bushmeat volume traded per inhabitant and settlement size; and (c) volume (in carcasses per annum) traded and the mean body mass of prey species in each site (data from Fa *et al.*, 2006).**

Bushmeat markets and measuring hunting sustainability

Observed differences in the volume of bushmeat traded may reflect the number of hunters operating, which in turn may be related to the population status of the prey species in the area (Fa *et al.*, 2005). However, as Ling and Milner-

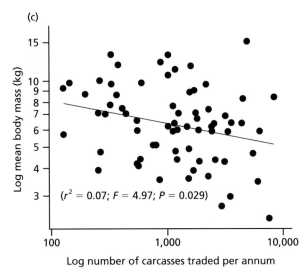

(c)

Log mean body mass (kg)

15

10
9
8
7

6

5

4

3   (r^2 = 0.07; F = 4.97; P = 0.029)

100   1,000   10,000

Log number of carcasses traded per annum

Figure 3.4   *Continued.*

Gulland (2006) argue, because open-access hunting is a dynamic system in which individual hunters respond to changes in hunting costs and prices obtained for their catch, resulting off-takes reflect human processes as well as prey abundance. Assessing underlying factors rather than proximate outcome variables is more complicated, but the trade-off in choosing to assess one or the other is between the potential for reduced monitoring frequency due to longer-term predictions and greater uncertainty through the introduction of additional assumptions (Ling and Milner-Gulland, 2006). Investing in characterizing supply and demand functions may not be essential if they are likely to change rapidly because of external economic or social processes or if effort and off-take can be manipulated directly. Assessing supply and demand, on the other hand, may be easier in a commercial market setting, because point demand is readily measured and elasticity of demand can be inferred from knowledge of cultural and economic conditions. Ling and Milner-Gulland (2006) suggest that to determine sustainability reliably some investment into modelling alternative monitoring and management strategies (with appropriate treatments of measurement error, system uncertainty and stochasticity), similar to those already being developed for fisheries, is necessary. Although this is an approach that definitely requires developing, its application may be more suited to small-scale analyses. In order to scale up to the level of large geographical areas, it may be necessary to sacrifice accuracy in order to gain a broader picture of impact of hunting on bushmeat species.

Bushmeat markets provide quantitative data on carcass numbers and price by species, and sometimes information on the origin of the meat can be obtained. Such data has been used to infer hunting sustainability. However, two major shortcomings of using market data have been discussed by Cowlishaw *et al.* (2005b). The first relates to the issue that, because most urban markets sell bushmeat from relatively large catchment areas, sustainability may then be incorrectly inferred from apparent market stability when the meat is simply coming from a previously untapped source. For example, Clayton and Milner-Gulland (2000) showed that, while wild pig sales and real prices showed no trends over a 10-year period in a market in Sulawesi, traders were in fact driving substantially further to obtain the meat. The other issue is that, as shown above, animals appearing in the market are only a selective proportion of the animals encountered when hunting. From the data presented here, hunters may be choosing to sell the larger-bodied animals, whilst low-value species tend to be eaten in camp or at home, given away or traded within the village, and hence never reach urban markets (Juste *et al.*, 1995; Rowcliffe *et al.*, 2003). Determinants of which animals are traded and which retained include the family's need for cash (de Merode *et al.*, 2004), the relative prices of bushmeat species and domestic meat (Wilkie and Godoy, 2001; Wilkie *et al.*, 2005) and transport costs to town (Crookes *et al.*, 2006). As well as changing the proportion of animals of different species traded and retained, increases in bushmeat prices should increase hunter effort, assuming that the bushmeat market is competitive, with no barriers to entry. If the resource is not depleted, increased effort should increase off-take (Damania *et al.*, 2005). Hence the effect of prices and costs on the volume of trade finally appearing in urban markets is complex.

Despite these caveats in measuring sustainability using market data, Crookes *et al.* (2006) argue that proxies can be employed to assess faunal depletion. Linked to the extinction filter effect, if a decline in mean biomass of individuals, or an increase in the mean intrinsic rate of increase (or its corollary, mean body mass) is detected, the inference can be made that more vulnerable, larger-bodied, slower-growing species are being lost. A secondary proxy is the use of an increase in the price, possibly coupled with a reduction in trade volumes, suggesting that animals are becoming scarcer and demand is not being met. Finally, faunal depletion can be inferred by a shift in the source of animals to less depleted areas, usually areas that are more costly to get to market owing to distance, road quality or terrain, or which have been newly opened by roads.

Data derived from the main market in Bioko Island, in Fa *et al.* (2000), have demonstrated relatively conclusively that over a period of 5 years (1991 to 1996), the profile of species appearing in the main city market (Malabo) shifted dramatically from larger to smaller species. As most hunted animals in the island are known to be sold to the Malabo market (Colell *et al.*, 1994; Reid *et al.*, 2005), even though data on hunter numbers were not available, the decline in larger species can be explained only by overhunting. Another study that employed off-take data, including information on species identity, capture location and sales price, from a survey of the Takoradi market, Ghana, concluded that the trade was sustainable as the result of a series of non-random extinctions from historical hunting (Cowlishaw *et al.*, 2005b). Vulnerable taxa (slow reproducers) have been depleted heavily in the past, so that only robust taxa (fast reproducers), such as rodents and small antelope, are now traded (Figure 3.5). Basically, because the intrinsic rate of species is correlated with the body mass, a drop in mean body mass of prey is a good indicator of faunal depletion, as already shown by Jerozolimski and Peres (2003). Cowlishaw *et al.* (2005b) suggest that Takoradi is a mature bushmeat market, where a sustainable level of trade has been reached. These authors also argue that the pattern of species depletion in Takoradi matched hunter perceptions of species vulnerability elsewhere in the country, where brush-tailed porcupine (*Atherurus africanus*) and Maxwell's duiker (*Cephalophus maxwelli*) are widely thought not to have been affected by hunting, in contrast to yellow-backed duiker (*C. sylvicultor*), giant hog (*Hylochoerus meinertzhagene*), red-capped mangabey (*Cercocebus torquatus*), and the African buffalo (*Syncerus caffer*), which have been severely depleted or driven to extinction locally.

The 'post-depletion profile' exemplified by the Takoradi city market occurs in other areas in West and Central Africa. Essentially, in such markets, species found are generally from forest-edge and farmbush, and many are also likely to be significant crop pests. These patterns are more likely to be typical of urban

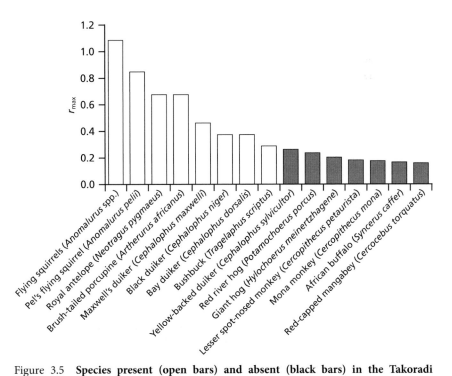

Figure 3.5 **Species present (open bars) and absent (black bars) in the Takoradi market, Ghana, arranged according to their intrinsic rate of increase (r_{max}). Data from Cowlishaw *et al.* (2005b).**

population centres, but a spectrum of market profiles, indicating from less to more exploited situations, emerges. Data from the Cross-Sanaga study (Fa *et al.*, 2006) can be used to verify this. Using the over 80 sites from this study, it is possible to show that mean body mass of prey is positively correlated with the proportion of ungulates in the market, and negatively with the proportion of rodents, both curves being highly significant (Figure 3.6). Furthermore, as shown in Figure 3.4c, higher volume markets are selling smaller-bodied species, as the mean body mass of prey is significantly negatively correlated with numbers of carcasses traded per annum. This means that the triangle, greater depletion → lower body mass prey → more number of carcasses in

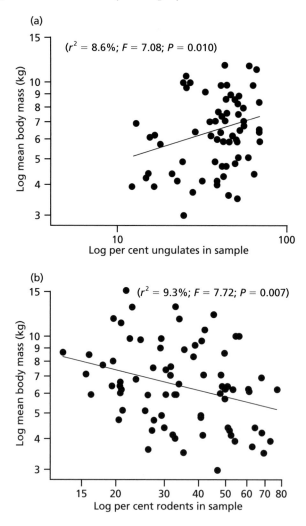

Figure 3.6 **Log–log relationship between the percentage of (a) ungulates and (b) rodents in sampled sites and the mean body mass (kg) in Cross-Sanaga sites (data from Fa *et al.*, 2006).**

markets, can be used to identify profiles that distinguish between a sustainable system at equilibrium and other systems at different stages of depletion. Additional information would help to further substantiate these findings, especially where more detailed data on the abundance and productivity of prey populations in market catchments and the scale of rural extraction can be gathered. In addition, further data on temporal changes to the size of the catchment area and the overall number of hunters in the system (both of which should remain constant under stable conditions), and on the costs and revenues for hunter trips (which should also remain equal when the system is stable), would be informative and relatively easy to collect. In fact, Crookes *et al.* (2006) produce evidence that suggests that such a dataset is of the quality that could realistically be aspired to by researchers and policy-makers wishing to use market data to carry out broad-scale sustainability assessments for the bushmeat trade. Further assessment of its strengths and limitations as a tool for sustainability analysis is required.

White elephant or red herring?

In this chapter, data have been presented on the characteristics of bushmeat markets, including some analyses on the information emerging from these to assess condition of wildlife in their surrounding areas. Hunting and bushmeat trade provide the main source of income for a large network of people ranging from hunters and farmers to market women and helpers in both urban and rural communities. In many areas the gathering and marketing of wildlife, even species such as snails that may be considered to be of little value, provides a significant proportion of the household cash income.

The reality is that wildlife populations in many parts of the African continent are declining as a result of overexploitation and destruction of wildlife habitat. Current levels of wild animal exploitation are not sustainable probably anywhere on the continent and areas where large populations of wildlife, as in southern Africa, still occur coincide with the enforcement of protection measures. From the data available from overexploited areas, it is reasonable to assume that, once large-bodied species have disappeared, the remaining species may be harvested sustainably. But, these 'post-depletion' scenarios should be avoided by instituting bushmeat management policies that protect the vulnerable species from hunting, but allow robust species to supply a sustainable trade (*see also* Wilkie, 2001). Clearly, such insights have without any doubt been acquired through investigations carried out of bushmeat markets. Should bushmeat markets still be studied, and is it cost-effective?

Knowledge and understanding of trends in bushmeat extraction is still fundamental to pre-empt any further erosion of African wildlife. However, monitoring wildlife (including dead wildlife as in markets) within large areas is problematic and costly, but necessary. It may be that carcass counts taken from bushmeat markets may not produce sufficiently robust correlational data on the population health of traded species, but this has not been sufficiently tested. For example, parallel studies of the population status of bushmeat species in forest and their levels in supply markets have not been undertaken. However, issues surrounding the mapping of hunting catchment areas (sometimes for numerous hunters) supplying a market, coupled with the methodological problems of accurately censusing animal populations, may make this unrealistic.

Given the relative ease of counting animal carcasses (albeit with the caveat that some protected species may be hidden from view) in markets, if large market numbers can be monitored [as already undertaken in Gabon by Wilkie *et al.* (2005) or Cross-Sanaga in Fa *et al.* (2006)], these often represent the best compromise between economy of collection effort and precision and accuracy of the estimates of population indexes. By standardizing data collection protocols and optimal sampling periods (as indicated in Fa *et al.*, 2004) it is possible to enable comparisons between areas and with other studies. The quality of the data, however, is important, and depends upon the continued dedication and training of the observers, the cooperation of various agencies and investigators and the rapid and accurate compilation of results. Such cooperative efforts, an example of which is the ongoing bushmeat market studies led by the Wildlife Conservation Society (WCS) and the government institutions in Gabon (WCS, 2006), is a likely way forward for understanding wildlife exploitation and use.

There is ample evidence that wildlife production is a feasible and viable form of land use in Africa, whether managed as wild populations in protected areas or in human-dominated habitats. The challenge for conservationists and wildlife managers is whether we can quickly elucidate the underlying principles of prey off-take and usefully infer the sustainability of hunting. As we have seen, such interpretations rely on the basic biological principle that species with slow life history strategies are more vulnerable to exploitation and that prey profiles lacking these species are therefore to some degree overexploited. Practical, well-gathered empirical data from markets, perhaps alongside applied modelling approaches [suggested by Rowcliffe *et al.* (2003) and Ling and Milner-Gulland (2006)] will formalize and clarify the condition of wildlife in large geographical areas, allowing snapshot prey profile data to indicate where in the spectrum of exploitation a given hunting system lies, and identify pockets of problems.

Notes

1 A white elephant is a valuable possession whose upkeep is excessively expensive, and may be useless apart from its value to the gifter and giftee.
2 Something that draws attention away from the central issue.

Cameroon: From Free Gift to Valued Commodity – the Bushmeat Commodity Chain Around the Dja Reserve

Hilary Solly

Introduction

This chapter looks at the bushmeat economy for the Bulu people living on the border of the Dja Reserve, Cameroon. It focuses on both the Bulu hunters and the local bushmeat traders, who purchase bushmeat from the hunters to sell in town. It describes how behaviour varies between different hunter 'types', as well as the role played by women as bushmeat traders. In addition, it points to the social and cultural as well as economic influences on those who hunt and trade in bushmeat.

The chapter is based on anthropological field research that took place between November 1996 and December 1998 on the borders of the Dja Reserve, Cameroon. The research was undertaken within the framework of APFT (Avenir des Peuples des Fôrets Tropicales), a 5-year research-based project funded by DG Development of the European Commission. The study was focused on the Bulu population, with the author living in Mekas, one of 20 Bulu hamlets and villages situated on the western border of the reserve. The Bulu are an ethnic group found throughout the tropical rainforest regions of southern Cameroon. They are often placed together with the Fang and Beti as making up the 'Pahouin' group, owing to their cultural, historical and linguistic similarities.

The Dja Reserve is renowned for its biological wealth and is recognized as a Biosphere Reserve and World Heritage Site by Unesco. It is located in south Cameroon's tropical forest region and covers a territory of 5,260 km² (Figure 4.1). The Dja River provides a natural boundary encircling approximately three-quarters of the protected area. The reserve is divided between two provinces, with 80% located in the East Province and 20% in the South Province. Since 1992 the South Province section has been managed by ECOFAC (Conservation et utilisation rationnelle des écosystèmes forestiers d'Afrique Centrale) in conjunction with the Cameroon Ministry of the Environment and Forests (MINEF).

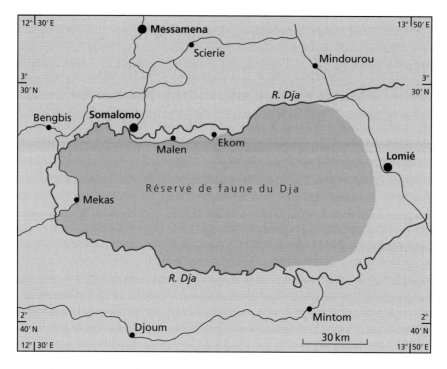

Figure 4.1 **Map of the Dja Reserve, Cameroon.**

The aim of the ECOFAC project is to combine biodiversity conservation with social and economic development for those populations dependent on the reserve's resources. One of the main concerns of the project is what it believes to be unsustainable levels of commercial bushmeat hunting in the Reserve (Delvingt, 1997). ECOFAC's conservation activities in the Dja Reserve have focused on two main areas:

- zoning, with the division of the reserve into two zones: a completely protected central zone, where access is forbidden, and a buffer zone to be sustainably managed as 'village forests' (with boundaries defining village territories);
- hunting control, using a combination of repression, with military-style *ecogardes* located in the villages and focused on stopping poaching activities; coherision, offering development opportunities to hunters in return for an acceptance by the local population of conservation activities within the reserve; and cooperation, encouraging a participatory approach, where the local population is involved in the sustainable management of the reserve.

The project's development activities have focused on:

- Substitution, by providing alternatives to those means of income that have been denied to the population. Alternatives for hunting income have included cocoa plantation regeneration, establishment of oil palm plantation, and employment with ECOFAC (as *ecogardes* and manual workers).

- Compensation, for the economic loss caused as a result of forbidding access to forest resources. Compensation activities have included improvements in education (building schools) and access (track improvement and bridge building).

Local Bulu perceptions of hunters and hunting

A questionnaire carried out in the village of Mekas to determine attitudes of 35 women (59% of the total female adult population in the village)[1] towards hunters and wildlife found that over half of women (54%) liked to be with men who hunted because it meant that they would have the opportunity to eat meat. The link was also made between bushmeat and money, with women mentioning how hunters were able to purchase petrol, soap and clothes so that the family could 'live well'. In addition, a hunter-husband was able to earn money during the 'dead season', that is the period of time before the harvest and sale of cocoa, when there are few opportunities for earning an income and money is short. Overall, 31% of women mentioned the ability to earn money as a positive characteristic of hunters.

However, not all women saw hunters in such a positive light: 3 of the 35 women questioned mentioned how hunters had money but did not work their farms. One woman said they were selfish and never gave their wives the game they killed. Another said she preferred a husband who farmed cocoa as income from game only came in small amounts, which women never get to see, whilst cocoa income was a larger lump sum used for the whole family.

Nearly all (34) of the women stated that forest animals were a problem for them as they destroyed the women's field crops (the cane rat, *Thryonomys swinderianus*, African brush-tailed porcupine, *Atherurus africanus*, and monkeys were cited as causing the most damage). Six of these women said that their husband or a relation set traps around their fields to reduce damage, and five women cleared the edge of their fields and built rough fencing to discourage animals. However, 16 women said that nothing was done to prevent these animals causing crop damage.

Men's views of hunting laid emphasis on the appeal of commercial bushmeat hunting in that the income gained is 'quick, easy money', involving low investment with rapid returns. A hunter need only purchase enough cable to lay a line of traps and take the time to place them and check them. In addition, rather than earning a one-off yearly income, as is the case for cocoa (the other main form of income for men), hunters are able to earn smaller amounts of money throughout the year. They also argued that income from hunting provided a vital stopgap during the dead season, with hunting being undertaken as a response to the unplanned and urgent need for money such as death or illness in the family.

However, it should also be noted that income from hunting was generally referred to by both men and women as being wasted or frittered away. There are a number of contributing factors relating to the 'easy earn, easy spend' nature of hunting income. First, Bulu culture shies away from the accumulation of individual wealth and towards gestures of generosity (Rowlands and Warnier, 1988; Geschiere, 1995). Individuals with money, particularly men, are under considerable pressure to display generosity. The small amounts of regular income from hunting are easily spent in a social setting of generosity, with the hunter buying alcohol (palm wine or distilled palm wine) and cigarettes for himself and his peers. Secondly, the Bulu regularly expressed the sentiment that nothing of value could be done with money earned from bushmeat. A typical response from a young man when asked what hunters do with the money earned from hunting was: 'It helps, for example, to buy (palm) wine or cigarettes … (but) money that can help him to buy clothes, he must work for in some other way.' Another time a hunter, when asked what he did with the money he earned from bushmeat said: 'I waste it, the money is not there to be saved'. He went on to explain that money earned from hunting was like money obtained through the use of witchcraft, with the witch often demanding the life of someone as a form of payment. Rather than earning money through hard labour, which is the case for income earned from cocoa farming, the hunter, like the witch, must make 'blood run' to get money quickly and easily. As with witchcraft, it was felt that nothing of value could be done with money earned in such a way.

Finally, there is the Bulu perception of the forest as being immense and containing inexhaustible wealth, including an unending supply of game (Solly, 2003). This, combined with the relative ease and low investment,[2] involved in hunting appears to lead to the view of game as a 'free gift' from the forest. This could well contribute to little value being given to the supply of bushmeat and subsequently little worth attributed to the income earned from it.

It should be noted that the opinions and views relating to hunters and hunting must be placed within the context of the presence of a conservation project that had been in the region for 7 years. Many of the local Bulu population were deeply resentful of the conservation activities of the project and what they perceived as its lack of development achievements. Questions posed relating to hunters and hunting were highly contentious and often resulted in anger and heated discussion. Even though the author was not a member of the ECOFAC project, her position was by no means neutral and this may well have influenced local responses to bushmeat-related questions.

The village-based bushmeat economy

The village-based bushmeat economy includes both hunters and bushmeat traders (local people who purchase bushmeat from hunters). Hunters were divided into different hunter 'types' according to their method and location of hunting activity. They were categorized according to whether they predominantly undertook 'village trap hunting', 'village gun hunting' or 'forest hunting'. Village hunting takes place within a days walking distance from the village (between 5 and 10 km). 'Village trap hunting' is largely undertaken around the hunter's own fields and plantations and is largely included as part of his daily farm chores. 'Village gun hunting' mainly occurs between dusk and dawn. A hunter has to own or borrow a gun and purchase cartridges. During the night he uses a head lamp to illuminate and startle the game before killing it. 'Forest hunting' involves spending at least one night away from the village in order to reach and return from the hunting location. Forest hunters travel up to 40 km from the village and can remain away for more than 2 weeks. These hunters have camps located in the forest with huts where they smoke and store the game before returning with it to the village. Most forest hunting was undertaken using snares, with some hunters also taking a gun with them.

The non-hunter bushmeat traders are men and women from the villages in the reserve who buy game from the hunters in order to sell on at a profit outside the reserve. They are generally able to sell the bushmeat just outside the reserve at double the village purchase price. If they travel into town to sell the bushmeat they can further increase their profit. However, most choose to sell their meat at the exit of the reserve.

Hunters and bushmeat

The activities of the 53[3] regular hunters from the village of Mekas were followed during a 1-year period between November 1997 and October 1998 (two hunters refused to participate).The hunters were aged between 15 and 65 years, with the highest number (32%) being between 31 and 35 years old. The study focused on bushmeat the hunters brought back to the village and what they did with this meat. The date and time of the hunter's departure and return were noted, as was the hunting technique used. When they returned, the hunters were asked which species they had killed and what they had done with the meat. If they had sold the meat then they were asked where it had been sold, e.g. to another villager in the village or to a *buyem sellum* who had come from outside the village, or if they had carried the meat out of the reserve in order to sell it. The price they had obtained for the sale of the meat was also recorded. If the meat had been given as a gift, then the hunter was asked to whom it had been given.

The data collected did not include information on any meat that had been consumed or sold whilst the hunter was away hunting. The forest hunters would consume some of their meat whilst away hunting. It is also possible that some bushmeat was sold to *buyem sellums* who visited the hunters at their camps in the forest or that some hunters took their kill directly out of the reserve to sell without first returning to the village. This is particularly likely in the case of category 'A' protected, or 'prize' game, such as elephant, gorilla, chimpanzee or leopard. Cameroon law strictly forbids the hunting of these animals and great care was taken to hide evidence of such activities.

Generally, there was a certain amount of sensitivity to questions linked to the hunting and selling of bushmeat owing to the illegal nature of the activity. The assistant employed to gather the hunting data was both a hunter himself and belonged to the same *mvôk* (family group) as the most serious group of hunters, which eased the collection of information. However, the question of how the hunters spent the money they had earned through hunting was stopped after a certain period because of the reluctant and sometimes aggressive response it received. It was felt necessary to drop the question in order to maintain good relations with the hunters.

In addition, it is likely that the person collecting the hunting data did not manage to record every hunting trip made, particularly in the case of those who set traps sporadically around their fields or lived on the outskirts of the village.

The results obtained (Table 4.1) indicate that forest hunters spent considerably more time away hunting than their fellow hunters, spending a total of 23,719 hours hunting (81% of the total time spent hunting by all groups of hunters). This compared with 4,099 hours (14%) spent by village trap hunters and 1,598 hours (5%) spent away by village gun hunters. In addition, it should be remembered that village trap hunters were likely to be undertaking farming activities whilst away checking or laying their traps. Overall, forest hunters earned most because of their numbers, but the average income per individual hunter was the greatest for village gun hunters (102,792 CFAfr for gun hunters compared with 72,432 CFAfr for forest hunters and only 28,244 CFAfr for village hunters; US$1 equivalent to 497 CFAfr). Village gun hunters are able to earn the most money owing to their efficient marketing methods. Although forest hunters sell a greater proportion of their meat than village gun hunters (74% compared with 60% for gun hunters and 56% for village trap hunters), they sell only 17% outside the reserve, compared with 74% for village gun hunters. This is probably because after a long period in the forest they return exhausted to the village and have no desire to walk the 14 km out of the reserve to sell their game. In contrast, village gun hunters are very efficient. They spend the least time hunting and then sell most of their meat outside the reserve, where greater profits can be made. Village trap hunters sell 90% of their bushmeat in the village but they are efficient with the time spent hunting in the sense that they visit their traps at the same time as undertaking farming activities. In addition, by hunting around their fields and plantations they are providing some protection from crop pests. Forest hunters, on the other hand, leave the village for long periods of time exclusively to hunt, and are therefore unable to undertake any other activity, effectively abandoning their farms and plantations whilst they are away.

Table 4.1 General characteristics of different hunter types. Exchange rate is US$1 to 497 CFAfr.

	No. of hunters	No. of hours away	Total earnings (CFAfr) per year	Earnings per person (CFAfr)	Animals killed		Animals sold		Animals given away		Where sold	
					No.	%	No.	%	No.	%	Mekas (%)	Outside reserve (%)
Village trap hunting	26	4099 (14%)	734,350 (28%)	28,244	720	41	404.5	56	315.5	44	90	10
Village gun hunting	10	1598 (5%)	616,750 (24%)	102,792	428	24.5	256.75	60	171.25	40	40	60
Forest hunting	17	23,719 (81%)	1,231,350 (48%)	72,432	602	34.5	445	74	157	26	83	17
Total	53	29,416 (100%)	2,582,450 (100%)		1,750	100	1,106.25	63	643.75	37		

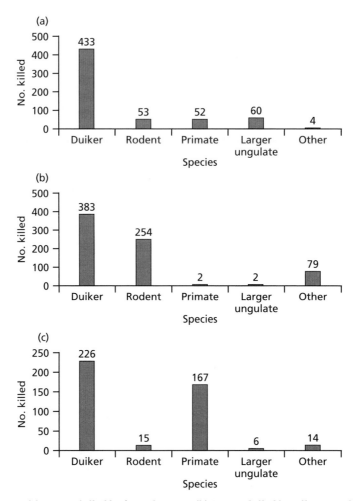

Figure 4.2   **(a) Species killed by forest hunters. (b) Species killed by village trap hunters. (c) Species killed by village gun hunters.**

Species killed also varied between the different hunter types, although the major species killed by all hunters was the blue duiker, *Cephalophus monticola*, making up 28% of the kill from forest hunting, 33% from village trap hunting and 44% from village gun hunting. Forest hunters took a greater proportion of duiker (71% of their catch) and larger ungulates (10% of their catch) than the other hunter types (Figure 4.2a). In addition, they were the only group to kill the red river hog (*Potamochoerus porcus*), as it tends to be found deeper in the forest. It also has a reputation for being aggressive and dangerous, requiring the skill and courage of these forest hunters to be killed. Forest hunters also killed a larger number of Peter's duiker (*C. callipygus*). This species is perceived to dominate hunting areas far from the village (Ngnegueu, 1998). Other species killed during forest hunting but which were not killed close to the village were the water chevrotain (*Hyemoschus aquaticus*), black and white colobus (*Colobus guereza*) and chimpanzee (*Pan troglodytes*). In contrast, village trap hunters killed far more rodents (35% of their catch; Figure 4.2b), dominated by the African brush-tailed porcupine (*Atherurus africanas*) and the Gambian giant rat (*Cricetomys gambianus*). Both species are attracted to crops in the farmers' fields and plantations, with *A. africanus* being cited as a major crop pest by women farmers. The main species killed by village gun hunters were duikers (53% of their catch) and primates (39% of their catch; Figure 4.2c). Of the primates, the main species killed were the greater white-nosed monkey (*Cercopithecus nictitans*), moustached monkey (*C. cephus*) and mona monkey (*C. mona*). The reason for the high number of primates killed is the ease with which they can be shot with a gun rather than in a trap or snare. Primates are also attracted to field crops and were cited by women farmers as being a major crop pest.

Non-hunter bushmeat traders

The vast majority of bushmeat is sold in the village to local people (81%). A key question is 'what is being done with this meat?' Is it for local consumption, or does this purchased game have an alternative destination? Data collected on goods being taken by local people out of the reserve suggest that a significant proportion of bushmeat bought by locals is destined for the urban market.

Data were collected over a 1-year period on all goods and people entering and leaving the reserve by ferry at its southern point. A ferryman transported people who wished to cross the Dja River at this point. He charged a small fee (normally 100 CFAfr) for the service. Between November 1997 and October 1998 every journey made by an individual, family or group was recorded by this ferryman. He asked each traveller or group travelling together where they were coming from and going to, their motive for the journey and what goods, if any, they had with them.

A total of 2,928 journeys were recorded, 1,417 journeys into the reserve and 1,511 journeys leaving the reserve. A total of 2,257 people travelled out of the reserve and 2,054 people entered or returned during this period of time. It should be noted, that although we were able to register both the outward and homeward bound journeys of many individuals, sometimes people would enter or leave the reserve from a different point. In these cases only one journey would be registered. In addition, it is possible that some people crossed the river using some other means than the ferryman (although there was no other ferry service). In this case their journey would not have been recorded.

The research results revealed that the major destination for trips out of the reserve was the town of Sangmelima, around 40 km from the southern border of the reserve. However, destinations also included the nearby towns of Mvomeka'a and Meyomessala, and the city of Yaoundé (approximately 200 km away), as well as the villages located along the roads to these urban centres.

The two principal reasons given for why people were making the trip out of the reserve were to sell bushmeat (20%) and to buy produce (20%). Forty-two per cent of trips out of the reserve involved the transport of bushmeat. An estimated 1,471 forest animals were carried out of the reserve during this period. Of the bushmeat taken out, 71% was to sell, 18% was for consumption and 8% to be given as a gift. Of those carrying bushmeat out of the reserve 63% were men, 32% women and 5% couples.

It is interesting to note that of those who left with bushmeat, 64% returned with town goods to sell, implying that they had used the income from the bushmeat to finance the purchase of town goods, with the specific aim of gaining further profit by selling these goods on in the village. By buying game from the hunters it seems that the local bushmeat traders transform the bushmeat from 'free gift' to valued commodity. Because of the value added through this exchange of money they feel the requirement to reinvest the income earned from its sale. Those who left the reserve with bushmeat were returning with goods worth nearly 28% of the total sales value of goods brought into the reserve during this period. With certain goods this increased to nearly 40% of the total sales value. In addition, the percentage of goods to be sold for commercial purposes was consistently, if not dramatically, higher for those who had carried bushmeat out of the reserve. These figures reveal the importance for both men and women of the commercial role that income from bushmeat sales outside the reserve plays in the local economy.

Conclusion

It is extremely important that organizations involved in biodiversity conservation and implicated in identifying development opportunities for local populations living close to protected areas have a detailed understanding of the social, economic and cultural specificities of these populations.

In the Dja Reserve, where the main concern of the ECOFAC conservation project is what it believes to be unsustainable levels of local and commercial bushmeat hunting, the focus must be on a detailed analysis of all those involved in the village-based bushmeat trade. It is important to discover if hunters are a homogeneous group or can be further divided into categories of hunting specialization. If this is the case then it is necessary to understand the motives and influences that affect the behaviour of these different types of hunter. It is also vital to search for and analyse local people who are not hunters but who are nevertheless involved in the purchase and sale of bushmeat.

One important issue is the contradictory nature of hunting income, perceived both as a vital 'stopgap' used in emergencies but also as a little-valued form of regular income, that is therefore wasted or frittered away. The need to understand the social and cultural, as well as economic, basis for this perception is vital for a conservation project aiming to work in close collaboration with the local people. In this case, the 'easy come, easy go' nature of bushmeat income is linked to its perceived inexhaustibility as a forest resource and its association with witchcraft. In addition, the fact that it generally provides a small but regular income means that it is easily spent on social gestures of generosity that fit comfortably within a culture that feels uneasy with the accumulation of individual wealth.

Knowledge about the differences between hunter types is extremely relevant to the project's conservation goals and policy. It should be noted that the majority of hunters (68%) hunt close to the village. These village hunters are providing a degree of protection for their crops from damage by animals attracted to their fields and plantations. They also are not entering deep into the reserve, where the ECOFAC project wishes to create a central, entirely protected zone. In addition, village gun hunters are efficient both with the time they spend hunting and at maximizing profits. Village trap hunting is largely subsistence in nature, with low time investment. All of these features could be incorporated into a conservation policy negotiated with the local people and permitting a certain level of both subsistence and commercial village hunting. In contrast, the characteristics of forest hunters are that they venture deep into the forest, invest a high level of time in exclusively hunting and generally obtain lower profit margins with their hunting activity than village gun hunters, who are more efficient in terms of time spent hunting and profit made from selling their catch. These characteristics suggest the development of a conservation policy that limits or forbids forest hunting, particularly in the central, protected zone of the reserve.

The research results also emphasize how important it is for project to focus its conservation and development policy not only on the activities of hunters, but further up the local commodity chain. In the Dja case this means gaining an understanding of the local bushmeat traders who purchase game from the hunters. One important feature of these traders is their transformation of bushmeat, through its purchase, from low-valued 'free gift' to highly valued commercial commodity. The important contribution that these bushmeat traders make to the local economy also needs to be acknowledged, as well as the role played by women. Again, this implies the incorporation of a certain level of commercial bushmeat hunting and trading into conservation policy, as well as the search for alternative income opportunities for these male and female bushmeat traders, rather than hunters alone.

Notes

1   Some of the women questioned were from the same household. There were a total of 32 households in the village.
2   It costs around 5,000 CFAfr (US$1 equivalent to 497 CFAfr) to make approximately 45 snares. Gun hunting involves the initial investment of purchasing a gun. There were six guns in the village, which were lent to hunters in return for a proportion

of the kill. Cartridges cost 500 CFAfr each. Building barriers around fields or in the forest involves time and energy but was rarely undertaken. Compared with the years of labour involved in establishing a cocoa plantation (up to 5 years before a cocoa tree is mature for harvest), hunting involves relatively low investment.

3 Seventy-four per cent of the male population, 66% of households (21 out of 32 households), earned income from hunting.

5

Determinants of Bushmeat Consumption and Trade in Continental Equatorial Guinea: an Urban–Rural Comparison

Noëlle F. Kümpel, Tamsyn East, Nick Keylock,
J. Marcus Rowcliffe, Guy Cowlishaw and
E. J. Milner-Gulland

Introduction

Humans have been hunting bushmeat for subsistence in the forests of equatorial Africa for millennia. However, the effective human population size in forested areas has increased dramatically in recent years as a result of population growth, habitat loss and improved access, and bushmeat consumption in many areas is now thought to be unsustainable (Milner-Gulland *et al.*, 2003). Why people consume bushmeat is not well understood, and generally varies between urban and rural areas and between economically disparate countries or regions. It has often been assumed that rural people are dependent on bushmeat for basic food supply (Ntiamoa-Baidu, 1998; Milner-Gulland *et al.*, 2003), but increasingly studies have found bushmeat to be more important as a source of income for rural communities, enabling them to participate in a market economy (de Merode *et al.*, 2004; Kümpel, 2006). This commercial trade depends on market demand for bushmeat, and increasing demand from urban populations is thought to be largely behind the increasingly unsustainable rate of bushmeat harvesting. The reasons for urban demand for bushmeat vary; in some cases it may be due to a lack of other acceptable or affordable alternatives, or bushmeat may be a luxury good for which a wealthy urban elite are willing to pay premium prices (Asibey and Child, 1991). Understanding the determinants of bushmeat consumption is a critical first step in designing effective policy to mitigate the effects of an unsustainable trade, from both a conservation and a development perspective.

Often, a preference for bushmeat over domestic meats or fish, as part of one's 'traditional heritage', is assumed. However such an assumption ignores the great variability in taste, availability and price within the 'bushmeat' category. Traditional or religious preferences and taboos can become eroded as cultural identities are homogenized and the basic need for protein increases

(Peres, 1990; Bowen-Jones, 1998). In addition, preferences may in reality be linked to availability or perceived cost of alternatives.

To date there have been very few empirical studies analysing the determinants of bushmeat consumption. A study by Anstey (1991) describes changing consumption patterns and preferences for bushmeat in Liberia: preference for bushmeat in general decreased as consumption frequency decreased, as did preference for individual species as they became rarer. These results suggest that preferences are not static but linked to consumption frequency, which in turn may be linked to price or availability. Conversely, Fa *et al.* (2002b) found distinct differences in the preferences for different bushmeat species between the Fang and Bubi ethnic groups on Bioko Island, Equatorial Guinea. For the Bubi, preference for and consumption of bushmeat species was linked to price and availability, as they generally could only afford the cheaper and widely available bushmeat species. There was a lack of correspondence between preferred and consumed meats for the Fang. The authors attributed this to the fact that the Fang, having originated from the mainland, have historically been exposed to many more species, and therefore retain preferences for foods they no longer consume. However, some Fang may still have been consuming their favoured species, either as imports from or on visits to continental Río Muni.

A shortcoming of many studies is that they consider consumption of and preferences for only bushmeat, and fail to include alternative protein sources. Recently, however, researchers in Gabon conducted two-choice taste tests, consisting of a popular type of bushmeat versus a popular type of domestic meat, on both urban and rural consumers (Schenck *et al.*, 2006). They found only a weak preference for bushmeat overall (even though the study sample overrepresented bushmeat consumers), and that only rural consumers had a consistent preference for bushmeat over domestic meat. They concluded that not simply taste but also price, familiarity, tradition and prestige drive the demand for bushmeat. They also found that consumers differentiate between bushmeat species, and therefore substitutes such as fish and chicken may vary in their acceptability depending on the species of bushmeat.

Changes in wealth and income are also likely to affect consumption patterns. The direction and magnitude of these changes depend particularly on the income elasticity of demand (Wilkie and Godoy, 2001; Apaza *et al.*, 2002). Wilkie *et al.* (2005) studied the role of prices and wealth in consumer demand for bushmeat in Gabon. They found that consumption of bushmeat, fish, chicken and livestock all increased with wealth and, where the price of these foods was higher, consumption was lower. This is the only study of its kind conducted in Africa to date (but see East *et al.*, 2005).

It is necessary to understand why people are eating bushmeat both to predict the effects of changing economics on consumer behaviour, and thus its implications for sustainability of the bushmeat trade, and to evaluate the effects of policies on potentially vulnerable actors in the commodity chain. If urban consumption is indeed driving hunting, we need to understand how it is doing so and whether it is targeting and threatening particular species and, if so, what policies might work to conserve wildlife without compromising food security, cultural tradition, national development or people's ability to meet basic needs.

We assess the relationship between consumption of different types of meat and fish, the price and availability of these products and the income and preferences of its consumers, making a comparison between the city of Bata and the village of Sendje in Río Muni, Equatorial Guinea. We also evaluate the dependence of Sendje villagers on bushmeat for their livelihood. We carried out surveys of households, bushmeat consumers and markets in Bata, and households, hunters and off-take in Sendje, in order to get a detailed understanding of the different determinants of bushmeat consumption. By comparing bushmeat consumption and preferences with those for other commodities, we place it within the context of available substitutes. We then ascertain future changes in consumption given likely economic development within the country, and how this may affect rural livelihoods, and suggest possible policy responses.

Study area

Río Muni covers an area of 28,000 km² and forms the continental part of the Republic of Equatorial Guinea (Figure 5.1). In the centre is the country's largest protected area, Monte Alén National Park, which covers an area of 2,000 km². In spite of its relatively small size, the park contains a wealth of biodiversity, including 105 species of mammal, of which 16 are primates (Gonzalez Kirchner, 1994; Garcia Yuste 1995). The majority of the population on Río Muni are of the Fang tribe (traditionally from the interior), but there are also significant numbers of other tribes, such as the coastal Ndowe, particularly in Bata, the regional capital.

Bata is starting to experience a boom in income and population, caused by an emerging oil industry (Pigeonniere, 2001; Energy Information Administration, 2003). The official monthly minimum wage increased from 25,000 CFAfr (US$46) in 2000 to 97,000 CFAfr (US$177) in 2003 (J. Ferreiro Villarino, personal communication) – more than tripling in 3 years, in real terms. The city of Bata has expanded rapidly in the last few years, with a population of 78,684 recorded in the 1994 National Census and 132,235 in the 2001 National Census (Ministério de Planificación y Desarrollo Económico, 2002).

The village of Sendje is a major source of bushmeat for Bata. It has about 300–350 inhabitants and is situated 41 km south of Bata and about 10 km west of the boundary of the 1998 Monte Mitra extension to Monte Alén National Park. In the past, there were human settlements and logging camps throughout the surrounding area of forest, but these are now abandoned and the only permanent settlements are along the main road. In the colonial era and for some years afterwards many people were employed in plantations (mainly

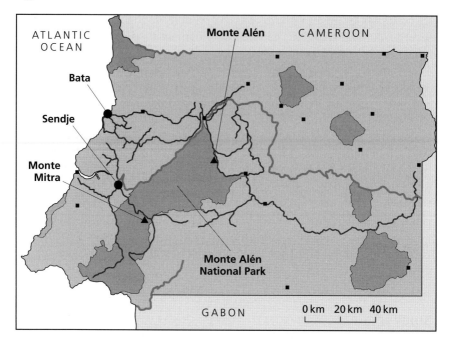

Figure 5.1 **Río Muni, with Bata, Sendje and Monte Alén National Park marked. Protected areas are shaded in dark grey.**

palm oil and coffee) and on logging concessions operating inside the present park but since independence political instability and economic decline have limited employment opportunities, and the majority of men in the village are now largely dependent on hunting for a living (Kümpel, 2006).

The bushmeat trade in Equatorial Guinea is little known, particularly in Río Muni. However, the trade is prevalent (Bowen-Jones, 1998), increasing and appears to be at least locally unsustainable (Fa *et al.*, 1995; Kümpel, 2006). There are still extensive tracts of forest in the country, sheltering relatively rich wildlife populations, including great apes, and very high densities of black colobus (*Colobus satanus*) inside Monte Alén National Park (Kümpel, 2006). The bushmeat trade is in theory regulated by official protection of particular species, but in practice laws are not enforced, and the trade is completely open regardless of a species' legal status.

Methods

Bata

In order to evaluate urban drivers of meat and fish consumption, interviews were conducted in Bata with a stratified, random sample of households and an opportunistic sample of fresh meat and fish purchasers in the central market (one of the two main markets in Bata) in May–July 2003 (see East *et al.*, 2005 for full methodology). A survey of fresh meat and fish on sale in the central market during 2003 provided information on availability and price (see Kümpel, 2006). Availability and price of all fresh and frozen food types is fairly constant across seasons, so this period is representative of market dynamics year-round.

One hundred households were sampled throughout the city, representing 867 people. Ninety-one per cent of these were Equatoguinean, of whom 73% were Fang, 25% Ndowe and the remaining 2% Bubi (from the island of Bioko, where Malabo, the capital of Equatorial Guinea, is situated). The interview was conducted with the household head together with the person responsible for preparing the meals. Interviewees were questioned on household structure and presence of selected wealth indicators and asked to name their first, second and third most consumed and most preferred meat and fish types. They were also asked about taboo foods, why they ate meat and if there were any foods they considered substitutes.

The market consumer survey focused on all the outlets in Bata's Central Market that sold fresh meat or fish. There were seven stalls selling predominantly bushmeat (as well as a limited number of locally sourced domestic animals such as poultry, pigs and goats), four selling fresh fish and one butcher's shop selling imported beef and goat. Each outlet was visited at different times on different days of the week. One hundred interviews with consumers were conducted at the bushmeat stalls and 30 each at the fresh fish and butcher's stalls, in each case immediately after they had made a purchase. The interviews were brief and included questions on interviewees' top three most consumed and most preferred meat and fish types, as for the household interviews.

The amount and price of each bushmeat species and availability and price of fresh domestic meat and fish on sale in the market was recorded daily during the study period (see Kümpel, 2006, for details).

We used a focus group to identify measures of wealth (see East *et al.*, 2005, for details). The group identified income as the major factor in consumption decisions, and type of employment as the key determinant of income. Household respondents were asked to indicate into which income category their total household earnings in the last month fell (with agreed indicators of income, such as lighting and cooking facilities, house type and vehicles, being used as a proxy for income for those respondents unwilling to indicate an income category). For market respondents, it was not feasible to obtain information on either income category or indicator assets, hence the occupation of the household head was recorded as a proxy for income (see Table 5.1 for a summary of head of household occupations for both the Bata and Sendje household samples).

Table 5.1   Occupation of household head (or main income contributor for the household) for Bata household, Bata market consumer and Sendje household samples, with percentage of occupations in each income category.

Occupation of household head	Bata market consumers (%)	Bata households (%)	Sendje households (%)
High income			
Accountant/banker/lawyer	6	4	0
Business manager/director	13	5	0
Civil servant	11	10	0
Doctor/nurse	3	3	0
Engineer/architect	2	5	0
Military/police	8	2	0
Office worker	2	2	0
Senior-level government official	2	4	0
Teacher/lecturer	4	6	0
Subtotal	*64*	*41*	*0*
Medium income			
Bar/restaurant/cook/maid	9	4	0
Construction worker	2	3	0
Driver	2	5	0
Factory worker	0	1	0
Manual labour	0	4	2
Religious	0	1	0
Skilled worker	7	21	5
Trader (small-scale)	5	7	3
Subtotal	*31*	*46*	*24*
Low income			
Housewife	0	4	0
Peasant	0	4	31
Student	1	1	0
Unemployed	3	3	0
Subtotal	*5*	*12*	*76*

Sendje

A study of rural consumption and preferences was conducted in Sendje in June 2002. Forty-one households (accounting for 288 people), out of a total of 56 in the village, were interviewed at random according to availability, ensuring that all areas of the village were covered equally. All people in Sendje were Fang, within seven different subtribes.

The majority of interviews were carried out with just one respondent (53%), who in most cases (86%) was the head of the household. Interviewees were asked about household structure, livelihood activities carried out by each member of the household and regularity of consumption (five categories: most days, weekly, monthly, rarely or never) and preference (five categories: love, like, neither like nor dislike, don't like or hate) for 37 different food types. As for the Bata households, interviewees were also asked about taboo foods and meat substitutes.

The price of all bushmeat and other meat and fish available in Sendje was recorded on a long-term basis during later fieldwork (November 2002 to January 2004: Kümpel, 2006). Hunter off-take data gathered from November 2002 to January 2004 were used to calculate the proportion of hunted animals traded and income earned from trade. Using data collected from repeat household interviews (conducted every 8 days with a sample of 42 households in Sendje from January 2003 to January 2004), income from hunting was compared with that from other livelihoods.

Results

Availability of meat and fish

Availability of meat and fish varied between commodities. In Bata, fresh beef (one cow every day or two, imported live from Cameroon on a monthly basis) and goat (two or three per day) were available every day from the butcher's shop in the Central Market, but often sold out by midday; this was the only source of fresh beef in the city until late 2003, when two other butchers shops opened. Fresh fish was available daily in the market except after heavy rains, but tended to sell out early in the morning; many people preferred to buy directly from the fishermen on the beach at dawn, who offered a cheaper price and greater choice.

Bushmeat arrived in the market throughout the morning, and some was usually still available in the afternoons. A wide range of bushmeat species were openly on sale (42 species), including those that were officially illegal to hunt or trade. Also available on the seven bushmeat stalls were pigs, goats, sheep, chickens and ducks, as well as freshwater fish, crayfish and crabs, the majority having been brought from villages outside Bata with the bushmeat, although the proportion of livestock and fish available was low compared with bushmeat. Many types of imported frozen meat and fish were widely available throughout Bata, most commonly mackerel and various cuts of chicken, pork and beef, from local shops and supermarkets as well as from 27 stalls in the central market.

In Sendje, although nearly every household owned chickens or ducks, and sometimes goats as well, they were generally reared only to sell at market or reserved for consumption on special occasions. Thirty-four out of the 41 households interviewed in 2002 contained men who hunted to some extent, and most households (men, women and children) did some freshwater fishing, so both bushmeat and fresh fish were widely available in the village, either from the household's own catch or through purchase of another hunter's surplus. Frozen mackerel was available most days (and sometimes also frozen livestock such as chicken) in Sendje itself or in the adjacent village.

Consumption and preference

For the Bata household respondents, the number of times each meat or fish type was rated top for consumption or preference was summed to give a score for each food type. There was a clear negative relationship between consumption and preference, with the most highly preferred types tending to have low consumption scores, and vice versa (Figure 5.2). The top five most preferred foods were all fresh fish or bushmeat species: red snapper (*Lutjanus campechanus*) was the most preferred food, followed by porcupine (*Atherurus africanus*) and then blue duiker (*Cephalophus monticola*). The top five most consumed foods were all frozen, with frozen mackerel first, closely followed by frozen chicken and then frozen pork.

A negative relationship between consumption and preference is also evident when consumption and preference scores are summed within broader categories (Figure 5.3). For both Bata (Figure 5.3a) and Sendje (Figure 5.3b), frozen fish and livestock were the least preferred but most consumed food types, while fresh fish and then bushmeat were the most preferred types, but were less frequently consumed. This pattern was less polarized in Sendje, with consumption patterns closer to preferences, suggesting that there were fewer

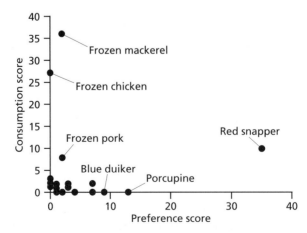

Figure 5.2   **First-choice urban consumption and preference scores for all types of meat and fish, based on the Bata household surveys. Highly preferred and consumed food types are indicated.**

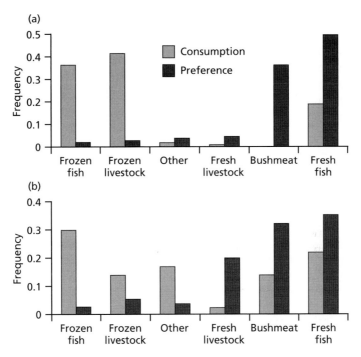

Figure 5.3 **Comparison of consumption and preference scores for different fresh and frozen food types, between (a) Bata and (b) Sendje households. Trends can be compared between the two samples, but not frequencies owing to differing methodology: for Bata, the sampling unit is the household, whilst for Sendje, it is the food type (see Kümpel 2006 for details). The 'other' category consisted of smoked, salted, dried and tinned foods; the large consumption score for this category in Sendje is due to the wide availability of cheap dried fish from the coastal town of Cogo on the Gabon border.**

constraints for rural consumers than for urban consumers. In both places, the distinction between food state is clearer than that between food type; frozen foods were more consumed, whereas fresh foods were by far the most preferred.

Reasons for the discrepancy between consumption and preference

Price variability between and within the different food types was much higher in Bata than in Sendje (Figure 5.4). At both urban and rural levels, fresh bushmeat and fresh livestock were the most expensive food types, and frozen fish the cheapest. Consumption of a food type was inversely correlated with price (Pearson's correlation; $R = -0.971$, $n = 5$, $P < 0.01$), showing that consumption decreased as the cost of a food increased. As fresh fish was only slightly more expensive than frozen foods in Bata, and actually cheaper than frozen livestock in Sendje, this may explain why it had a higher consumption

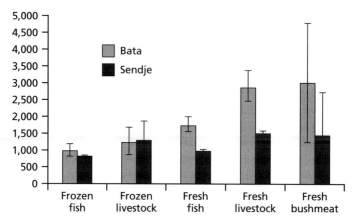

Figure 5.4 **Comparison of price variability for different meat and fish categories between Bata and Sendje. Prices are the 0.05–0.95 percentiles of a representative group of food types in each category. The 'other' category is not included here as it includes smoked, dried and tinned foods, which are difficult to compare directly with fresh or frozen cuts in terms of price per kilogram.**

score than the other fresh food types, and therefore was closer to its preference score. Additionally, in Sendje, as many households fished, their catch cost nothing except overheads, so price was not such an important factor. The more extreme price differences in Bata, coupled with a greater disparity in availability of fresh and frozen food types, may explain the greater differences between consumption and preference scores than in Sendje.

In both samples there was great variation in prices within the bushmeat category. This partly explains the greater consumption of bushmeat in Sendje. The majority of households in Sendje hunted, and whilst they sold the higher-priced and more marketable species such as crocodile, pangolin and porcupine, cheaper species, and carcasses too small or rotten to sell, were frequently kept for their own consumption.

Specific consumption decisions

Meat and fish are important components of the Equatoguinean diet, and the majority of households in both Bata and Sendje said that they ate meat or fish every day. Both urban and rural households stated 'health' as the reason for eating meat or fish. People often used the term *'comida'* (food) to imply meat, and the idea of a balanced vegetarian diet was generally incomprehensible. Leafy greens were most commonly cited in both Bata and Sendje as the main substitute when meat or fish was not available. Protein-rich substitutes such as eggs, pulses and oily vegetables did not feature highly as meat substitutes, particularly in Sendje, where the availability and affordability of many of these substitutes was particularly low.

Having made the decision to eat meat or fish, the choice is then between meat or fish types. As detailed above, the distinction between eating fresh or frozen food appeared to be more important to consumers than that between livestock, bushmeat or fish. However, within food types, species-specific preferences were expressed, particularly with respect to bushmeat. Many people had taboo foods at the tribal, subtribal, family or individual level, with a particularly high number of taboos for women of child-bearing age and children. In Sendje the top three taboo species were gorilla (*Gorilla gorilla*), chimpanzee (*Pan troglodytes*), then snake, and in Bata, snake, monkey and chimpanzee (with gorilla a close fourth). Reasons for avoidance were varied, but included tradition, similarity to humans (for primates), an evil expression and fear of infertility (depending on the respondent's age, sex and reproductive state). The lack of preference for apes is reflected in the fact that they were the very cheapest meats per kilogram in both the market and the village.

For 12 bushmeat species, there were data available on both household consumption and preference in Bata, as well as prices and availability in the Central Market (Table 5.2). Linear regression analysis suggested that consumption was positively related to preference (adjusted $R^2 = 0.67$, $P < 0.005$), and availability (adjusted $R^2 = 0.48$, $P < 0.05$), although this did not have a significant effect independent of preference, which was also related to availability (adjusted $R^2 = 0.54$, $P < 0.01$). There were no trends in either price or species body mass. Hence, within the bushmeat food type, more common species appeared to be preferred, as well as consumed more often, regardless of price or species size.

Similarly, purchasers approached in the Central Market tended to give 'preference' or 'taste' as reasons for consuming a particular species, rather than price (Figure 5.5). It seems that, although consumers chose between broad food types on the basis of price, within food types choices were affected by complex interactions between tastes and availability, rather than by cost.

Major determinants of urban consumption

In Bata, income was a primary determinant of consumption frequency for meat and fish. In a multiple generalized linear model of household consumption against income and nationality, bushmeat, fresh livestock and fresh fish all showed significant positive coefficients for income, while frozen livestock and fish showed non-significant or negative coefficients (Table 5.3). In other words, consumption of fresh food types increased with increasing income, while consumption of frozen produce tended to decline (see East *et al.*, 2005, for more details). This result is backed up by the finding that purchasers of fresh produce in the market were a comparatively rich subset of the Bata population as a whole (Table 5.1).

The influence of ethnic origin, city district, educational level of the household head and religion were also tested for, but only the first of these

Table 5.2 Information on consumption, preference, availability and price of the 12 main bushmeat food types mentioned in the Bata household surveys, listed in decreasing order of consumption. Species are fresh unless otherwise stated. Average mass, availability (numbers sold) and prices are from Kümpel (2006) and consumption and preference scores from East et al. (2005).

Species (common English name)	Species (Latin name)	Adult body mass (kg)	Consumption score	Preference score	No. in market during 2003	Dressed price/ kg (CFAfr)	Price/carcass (CFAfr)
Blue duiker	Cephalophus monticola	4.8	0.297	0.087	3,034	2,022	6,344
Brush-tailed porcupine	Atherurus africanus	3.2	0.264	0.154	2,008	3,711	7,625
Tree pangolin	Phataginus tricuspis	2.0	0.093	0.084	435	5,414	7,071
Mandrill	Mandrillus sphinx	13.8	0.055	0.018	389	3,886	34,961
Putty-nosed monkey	Cercopithecus nictitans	5.2	0.044	0.01	1,205	2,093	7,000
Red duiker	Cephalophus spp.	20.4	0.039	0.009	738	2,919	34,181
Giant pouched rat	Cricetomys emini	1.2	0.028	0.01	81	3,306	2,500
Red river hog	Potamochoerus porcus	22.7	0.022	0.005	116	3,589	53,000
Cane rat	Thryonomys swinderianus	4.6	0.016	0.005	21	3,100	9,286
Hinge-backed tortoise	Kinixis erosa	1.7	0.011	0.01	1,189	2,903	3,200
Northern talapoin	Miopithecus ogouensis	1.2	0.011	0	26	2,121	2,545

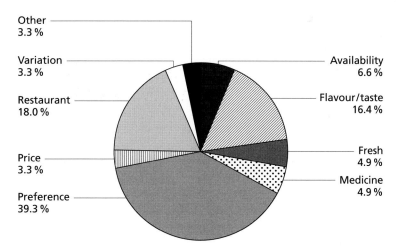

Figure 5.5 **Reasons cited by purchasers at bushmeat stalls in Bata's Central Market for buying a particular piece of bushmeat.**

had any significant effects on consumption (Table 5.3). Equatoguineans were much more likely than other nationalities to buy bushmeat, whereas purchasers of fresh domestic meat were more likely to be from other African countries (mainly Cameroon or Nigeria). The nationality of interviewees purchasing fish was more representative of the sample as a whole (Figure 5.6a). Among Equatoguineans, the Fang bought more bushmeat, while other tribes (predominantly the coastal Ndowe) bought more fish and to some extent domestic meat (Figure 5.6b).

Another important determinant of consumption patterns is one that is commonly overlooked and difficult to quantify: nutritional awareness or 'marketing' of foods and their relative benefits. Although cultural traditions were important (as is clear from the finding that ethnicity is an important determinant of consumption), there was a clear readiness of Equatoguineans to take on foreign ideas, culture and tastes. However, response to external influence was not always positive: there was a general perception that frozen produce has low nutritional value compared with fresh, and in Bata several respondents alleged that frozen food was dangerous as it had been dead a long time.

Dependence of rural people on hunting

The village off-take survey revealed that Sendje hunters sold the majority (mean 89%) of their catch. However, this proportion was more variable for hunters who caught fewer animals (with a significant proportion of animals being consumed by the household) and was more consistently higher for more prolific hunters (who were hunting nearly entirely for the commercial trade)

Table 5.3 **Results of generalized linear models of the factors affecting consumption rate, based on household surveys in Bata (see East *et al.*, 2005 for details). The coefficients for income are given, together with tribe/nationality if significant. Other factors were non-significant. Non-significant coefficients are in parentheses.**

Food type	Recall period	Explanatory variable	F	d.f.	P	Coefficients			
						Income	Fang	Ndowe	Other
Bushmeat	Day	Income	2.24	1, 97	0.14	(0.55)			
Bushmeat	Month	Income	8.23	1, 89	0.005	0.26			
		Tribe/nationality	6.81	2, 89	0.002		-1.51	-2.43	-2.41
Fresh livestock	Day	Income	4.51	1, 95	0.036	0.81			
		Tribe/nationality	4.50	2, 95	0.014		-13.14	-20.08	-11.32
Fresh fish	Day	Income	4.25	1, 97	0.042	0.55			
Frozen livestock	Day	Income	1.75	1, 97	0.19	(0.11)			
Frozen fish	Day	Income	4.10	1, 97	0.046	-0.27			

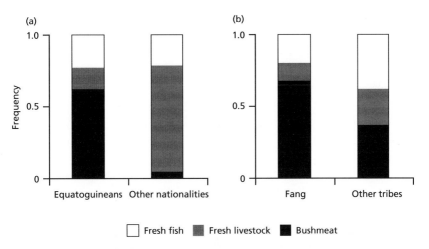

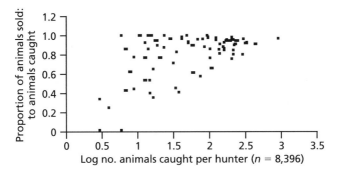

Figure 5.6 **Frequency of customers buying different fresh meat or fish types in Bata's central market, split (a) by nationality and (b) among Equatoguineans, by tribe. Chi-squared tests for significant departure from random: (a) $\chi^2 = 65.35$, d.f. = 2, $P < 0.01$; (b) $\chi^2 = 36.40$, d.f. = 2, $P < 0.01$.**

Figure 5.7 **Log$_{10}$ number of animals caught by proportion traded per hunter, from Sendje off-take survey during 2003 ($n = 8,396$ animals).**

(Figure 5.7). Given the higher price of bushmeat compared with alternatives in Sendje (Figure 5.4b), it made economic sense for hunters to sell their bushmeat and buy cheaper substitutes such as frozen mackerel to eat, using the profit to purchase necessities such as kerosene and soap and luxuries such as alcohol and cigarettes.

The long-term household survey showed that, at the village level, hunting was a major income-generating activity, falling just below waged employment and above bar trade, with agriculture and fishing of relatively minor importance (Figure 5.8). Hunting and waged employment were the main male livelihoods in the village, but men earned far less on average per month from hunting

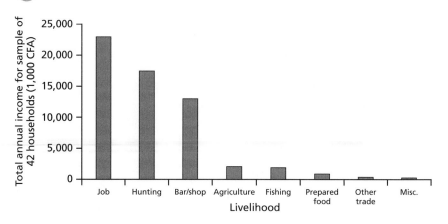

Figure 5.8   **Total income earned from different livelihood activities for the household sample in Sendje during 2003 (*n* = 1,607 interviews).**

(mean = 43,043 CFAfr; range = 0–195,678 CFAfr; *n* = 83) than from waged employment (mean = 88,394 CFAfr; range = 20,000–160,000 CFAfr; *n* = 30).

Discussion

Bata's consumers had a strong preference for fresh meat and fish over frozen, but on cost grounds they most often ate frozen foods. The degree to which they were able to satisfy their preferences was significantly related to their income. In this city, it seems that frozen produce is an inferior good, while fresh produce, including bushmeat, is a normal good. The perception that bushmeat is a luxury good in urban settings is commonly cited in the bushmeat literature (e.g. Bowen-Jones *et al.*, 2002). However, it seems that here fresh produce in general is the focus of demand, and at present has necessity (income coefficient < 1) rather than luxury (income coefficient > 1) status.

The most preferred food category was fresh fish, followed by bushmeat and then fresh domestic meat. There were particular preferred items within these categories, including red snapper amongst the fish species, and porcupine, blue duiker and pangolin (*Phataginus tricuspis*) amongst the bushmeat species. Other studies in Central Africa have shown similar orders of preference among bushmeat species (e.g. Njiforti, 1996). These species were also amongst the most readily available items in the market. Preference and availability were correlated in our analysis of individual bushmeat species, and it is likely that there is strong feedback between them.

Preference patterns were mirrored at the rural level in Sendje, as were consumption patterns to some extent. The reason for such similarities can be explained by Sendje's cultural homogeneity with, and geographical proximity to, Bata. The observed differences help to explain the determinants of bushmeat consumption. Where availability and accessibility to fresh foods is not such a constraint, as in Sendje, people consume more fresh fish and bushmeat, and less frozen fish, even though they are poorer.

Other studies of bushmeat consumption have not observed widespread use of frozen foods as a cheap source of protein, or often even considered the interaction between fresh and frozen food consumption. Two-choice taste preference tests in Gabon demonstrated that domestic meat may be an acceptable substitute for bushmeat there (Schenck *et al.*, 2006). However, when asked about stated preferences, only 28% of respondents stated fish compared with 44% for bushmeat, suggesting that fish is less universally preferred in Gabon than in our study. The authors warned that increasing the availability of the low-quality beef typically sold in Gabon may not encourage a switch away from eating bushmeat, but made no mention of the general acceptability of frozen meat and fish compared with fresh. Our finding that 'state' (which may be a proxy for real or perceived quality) is more important than 'type' has important implications for management, as at present Equatorial Guinea relies heavily on imports of frozen produce as a supply of animal protein.

Currently, bushmeat is widely available and consumption is constrained primarily by income (with ethnicity and, to some extent, cultural taboos also a factor). This is backed up by the fact that purchasers of bushmeat in the market were a richer subset of the population as a whole and restaurant clients eating bushmeat were predominantly male workers. These findings suggest that bushmeat consumption is likely to increase rapidly as Bata's economic prosperity and population continue to grow. On the other hand, it is possible that this trend may be countered by a Westernization of tastes associated with economic development. For example, specific bushmeat preferences in Río Muni have shifted since the study of Sabater-Pi and Groves (1972) (particularly away from mandrills, *Mandrillus sphinx*), and our sample included a number of respondents who stated that they did not eat bushmeat because it was 'dirty meat'.

In Gabon, fish was found to be a substitute for bushmeat (Wilkie *et al.*, 2005). This is also almost certainly the case in Equatorial Guinea, given the particularly strong preference expressed for fresh fish. The FAO has set up a project in Equatorial Guinea aimed at expanding the capacity of sustainable fisheries (FAO, 2003). However, recent linkages between reduction in supply of marine fish stocks and bushmeat exploitation in Ghana (Brashares *et al.*, 2004) are a warning that, unless coordinated management of bushmeat and fish is undertaken, protection of one may result in depletion of the other.

Freshwater fish supplies are not much better off: in sub-Saharan Africa, annual per capita freshwater fish supply dropped from 9 kg in 1973 to 6.6 kg in 2001, compared with an increase from 12 to 16 kg worldwide (von Bubnoff, 2005). In contrast to bushmeat, very little fish is traded to Bata from rural villages, although whether this is a result of dwindling supplies or low market demand relative to bushmeat is not entirely clear. There have been recent calls for urgent investment in aquaculture projects across Africa in order to match population growth and the subsequent demand for animal protein (von Bubnoff, 2005); apart from anecdotal reports of microscale aquaculture projects in Río Muni (S. Allebone-Webb, personal communication), this seems some way in the future for Equatorial Guinea.

Ultimately, if the sustainability of bushmeat and fresh fish off-take rates is to be ensured, the potential for substitute goods will need to be explored. A. L. Rose (2001) suggests that marketing of bushmeat alternatives has the potential to change consumption behaviour. Equatoguineans spend a relatively large proportion of their expendable income on heavily marketed and expensive Western products such as Coca-Cola or condensed milk as baby feed (N. Kümpel, personal observation). In theory, given sufficient economic incentive, awareness could be improved of the nutritional value of frozen meat and fish, or even vegetarian alternatives. Preferences can develop only once a product is reasonably familiar (Turrell, 1998).

Livestock rearing in Equatorial Guinea is vastly underdeveloped, and the results of our analyses of availability, consumption and preference all reflect this. Fresh domestic livestock currently scores low in terms of preferences in Bata. However, this does not necessarily mean that domestic meats could not become a more important component of the Equatoguinean diet. Fresh domestic livestock meat is scarce in Bata. Goats, sheep, pigs and poultry are reared in villages with no formal livestock husbandry or veterinary provision, and few of these animals reach the cities. The lack of cattle rearing in the country is attributed to the presence of tsetse fly (Ministry of Agriculture, Republic of Equatorial Guinea, personal communication), which transmits trypanosomiasis (although Equatorial Guinea only has low endemicity of this disease; WHO, 2004). Thus, beef in Bata comes from cows imported from northern Cameroon, but in 2003 these supplied only a total of three butcher's stalls in the entire city. However, neighbouring parts of southern Cameroon have an established livestock industry (Teale, 2003), including poultry and other trypanosomiasis-resistant species. Thus, it appears that, with political will, livestock rearing could be promoted more vigorously as a response to anticipated future demand for fresh produce. This could also present alternative livelihoods to hunting for rural people, many of whom presently depend on the bushmeat trade as a sole source of income. Controlled agricultural development, limited to the already degraded habitat immediately surrounding present villages, could be one component of a strategy to ensure the preservation of bushmeat hunting and biodiversity in Río Muni for future generations.

However, such initiatives will require considerable, expensive, long-term political support both in-country and from international development organizations. In the short term, both livelihoods and biodiversity could be preserved by effective regulation of the trade. There is little specific demand for threatened species most at risk of overexploitation, such as the great apes and black colobus. By allowing hunting outside protected areas and trade of only the more robust, fast-reproducing species, such as blue duiker and brush-tailed porcupine (28% and 26% respectively of the total off-take in Sendje during 2003), a sustainable trade could continue to contribute to livelihoods and national protein needs. This would not be straightforward to implement either (particularly given current incentives and capacity to enforce wildlife legislation), but with sufficient political, institutional and financial backing it could be a more immediate solution to the problem of overexploitation for bushmeat in Equatorial Guinea.

Acknowledgements

We thank the Ministry of Forestry and Infrastructure, INDEFOR and ECOFAC in Equatorial Guinea for supporting our research and Michael Allen for his help and advice. We are grateful to Cristina Ojoba Isambo, Rigoberto Esono Anvene, Bienvenido Ondo Ndong, Pedro Nsue Nseng and Antonio Ayong Nguema for their assistance and for sharing their knowledge with us. The study was funded by the UK's Natural Environment and Economic and Social Research Councils and Conservation International through the CARPE programme.

6

Livelihoods, Hunting and the Game Meat Trade in Northern Zambia

Taylor Brown and Stuart A. Marks

Introduction

This chapter examines the character, extent, costs and benefits of hunting and the game meat trade in villages adjacent to Zambia's Luangwa Valley national parks. It is essential to understand the role that game meat plays in local livelihoods if new conservation initiatives there are to be effective, if locals are to secure sufficient and sustainable alternative livelihoods and if illegal hunting in the national parks is to be stemmed.[1]

This chapter begins by examining the practice of hunting in the area: How are hunts organized? Where do they take place? Which types of animals are pursued and how are they killed? How is meat distributed? The study then examines what happens to game meat once hunters return to their home villages. We analyse the purchase of game meat by local and outside traders and its resale in Lusaka, the Copperbelt and other urban areas. The study then explores the costs and benefits of the game meat trade for locals in both quantitative and qualitative terms.

Overall, the study demonstrates that since the late 1970s an increased market demand for game meat and a lack of alternative sources of income has transformed the character of hunting. The study shows that illegal hunting and game meat sales are the most significant sources of cash income for many people in the study area. Perhaps one-third of all local households gain at least some income from this trade.

The chapter concludes by discussing the implications of the game meat trade for conservation initiatives and their possibilities for introducing alternative livelihood strategies to hunting.

Methods

This study is based on ethnographic fieldwork, interviews and case studies in several large, dispersed settlement clusters on the Central African plateau to the west of the Luangwa Valley. The origins of this study lie in 1973, when one of the authors interviewed a dozen hunters from the study site (see Marks, 1976; 2005). In 2001, as a consultant, he returned to the plateau and conducted interviews with another 30 hunters. This fieldwork and consultancy provide a baseline for assessing changes in hunting practices and for the quotes of interviewees at the beginning of the sections of this chapter.

In 2002, the authors employed a literate and well-connected local elder to conduct further interviews with hunters and traders.[2] As a respected local, this researcher knew his fellow villagers sufficiently to gain their trust and to judge the veracity of their claims. As part of this research, he interviewed and collected the life histories of an additional 26 long-term hunters and interviewed 10 regular game meat traders. He also drafted detailed accounts of 18 hunting expeditions, based on extended interviews with hunters and carriers fresh from their forays into the Luangwa Valley.[3]

The study has several limitations, although we believe that it presents an accurate portrait of hunting and the game meat trade in this part of Zambia. First, it relies on interviews rather than intensive participant observation to assess the character and extent of hunting and the game meat trade in the area. Second, there is an inherent selection bias in the choice of hunters interviewed; it is not a random sample, but reflects the hunters and traders known and trusted by the interviewee or available to the consultant. Third, the interviews represent a snapshot of specific hunts within a specific time frame (from September 2002 to February, 2003). Hunts were therefore assessed during both the dry season (when most large-scale hunts occur) and the rainy season (when hunting is more infrequent and usually of a smaller scale). This range goes some way to reducing seasonal bias within the dataset, but it is recognized that the estimates of the income generated by hunting would have benefited from a broader sample of hunts across an entire year. Finally, given the time and terms for the study, we worked on hunting processes and production and did not conduct household surveys.

The changing character of hunting

> The difference between now and in the past in terms of animal populations is that there is not much wildlife to see on the plateau, but there is in the valley. In the valley, it is not easy to hunt because of game guards, distance, and floods. We just take the risks.   Man, 40 years old, 9th-grade education

Hunting has long been a cornerstone of local livelihoods and identities. The Bisa people living in the study site historically relied on hunting as well as fixed and shifting cultivation, trade and raiding to secure a livelihood (A. D. Roberts, 1973; Marks, 2004). Hunting is socioculturally significant and plays a prominent role in Bisa social relations, kinship and ritual (Marks, 1976; 2005).

In the past several decades, however, the character of hunting in the study site has been transformed. Distant markets – not local needs or social systems

– are now the driving force for hunting on the plateau. As local hunters have sought to profit from hunting, they have expanded their hunting ranges to include distant parts of the Luangwa Valley, increasingly relied on guns to kill game, expanded the size of their hunting expeditions and altered their quarry preferences.

Until several decades ago, plateau villagers seldom ventured into the Luangwa Valley for their hunts: there was sufficient wildlife in the vicinity of their villages to satisfy their subsistence or trade needs. In the past several decades, however, as plateau game stocks have declined and demand for wildlife products has increased, hunters have been drawn progressively to hunt the large mammals in the valley. Of the hunt expeditions recounted for this study, all but one descended the escarpment into the Luangwa Valley and its national parks.[4]

The ivory and rhinoceros poaching 'boom' of the 1970s and 1980s first drew plateau hunters into the Valley to hunt on a regular basis (Leader-Williams, 1988; Leader-Williams and Milner-Gulland, 1993). These 'commercial' hunts also set the precedent for the recent profitable, large-scale and long-distance ones. By the 1990s, the supply and demand for ivory and rhinoceros horn had declined. At the same time, however, the market for game meat in Zambia's cities continued to grow.

As locals shifted from subsistence and small-scale local trade to larger-scale commercial trade in ivory and game meat, the size of hunting expeditions increased. Initially, most hunts on the plateau were small, involving a hunter and a few apprentices or carriers. Currently, however, most hunts comprise two or more hunters and engage 12–20 carriers. The larger the size of the hunting party, the greater the profit for the organizing hunter – each carrier is responsible for transporting at least one piece of meat for the hunter (including ivory if an elephant is killed). The more carriers engaged, the more meat the hunter is able to keep for himself.

Market demand for game meat has also transformed the types of game pursued by plateau hunters. Although one hunter (72 years old) recalled killing elephants, buffalo, hippo and eland on the plateau, the main quarry of local hunters there was typically small to medium-sized antelopes and wild pigs. When these mammals were killed or snared, the meat was distributed to residents within the hunter's village. Currently, however, the primary selection criterion for market hunters is the size of animal – shooting large animals enables hunters to maximize their carrier's loads while minimizing their time in the bush and exposure to detection by wildlife scouts (Milner-Gulland and Leader-Williams, 1992). As a result, elephants (now mostly females and young), hippos and buffalos are increasingly the preferred quarries, although species such as roan, zebra, warthog, duiker and other species are shot as well. Generally, warthog, duikers and small mammals are shot during a trip and consumed as provisions. Lions and hyenas are occasionally shot if they become a menace while the meat is prepared in the bush.

Contemporary hunting

In place of farming, I would like to have a job for pay as sometimes farming is not productive and this becomes difficult to get inputs and food for the family. This time animals are very few (around the villages) compared with the past. Now as a sign of depletion, hunters have to leave the village for about 6 days' walk before they can find some animals to kill – especially big animals. That hunting is possible only in the Luangwa Valley.   Man, 41 years old, 7th-grade education

It is difficult to gauge precisely the number of individuals and households benefiting from the game meat trade. Our research, however, suggests that in a population of approximately 2,600 around 75 men are hunters, i.e. they possess a gun, loan it or make forays into the bush, and derive at least some of their cash income from the game meat trade. There are more carriers than hunters. We estimate that between 170 and 225 men are regular carriers. Overall, hunting contributes to the livelihoods of perhaps one-third of households.

Hunters range in age from their early 20s to their 70s; the youngest hunter interviewed was 21, the oldest 72. Hunters tend to learn their bush skills from their matrilineal uncles, fathers and brothers-in-law. In most cases, expectant hunters apprentice as carriers with these relatives several years before organizing their own hunts. Hunting is a male-dominated activity: there are no female hunters or carriers in the area. Most hunters are married with children and other dependants. Carriers tend to be younger (in their early 20s or younger) and unmarried. Many of these young men have never completed primary school, and few seem to have alternative livelihood options.

Most hunting in the study site is carried out with locally manufactured muzzle-loading guns (*mfuti*). These guns are relatively inexpensive and readily replaced should they be confiscated by wildlife officials. The materials for a gun cost between 20,000 kwacha and K25,000 (about US$4.40–5.50), and a finished, assembled muzzle-loader costs around K50,000 about (US$11).[5] Although some hunters manufacture their own guns, most are made by one of half a dozen local gunsmiths. Muzzle-loaders are surprisingly effective – they can kill the largest of game, including elephants and hippos. Moreover, the ingredients for muzzle-loading ammunition are inexpensive and easily acquired. Gunpowder is locally manufactured from a combination of 'C-compound' (a fertilizer that contains saltpetre and potassium nitrate and is used primarily for tobacco farming) and the charcoal ashes of a common shrub, the 'ububa' (*Tephrosia vogeli*). A cupful of C-compound costs around K10,000 (about US$2.20) and provides enough saltpetre to make the gun powder to propel around 20 rounds of ammunition. The lead used as ammunition is derived either from dead car batteries or from the lead mines in the Copperbelt. A piece of lead large enough to make 8–10 bullets sells locally for about K3,000 (about US$0.65).

Locals own few rifles and no automatic weapons. A few hunters own old Arab-style muskets, but these are seldom used. Occasionally, a wealthy trader or government official will lend a rifle or an AK47 and ammunition to a hunter. They will then return several weeks later to collect the gun, unused ammunition and a percentage of the game meat or cash.

Local hunters may also snare game around their villages and near their temporary bush camps. Wire snares are readily available and are used for wild pigs and most antelopes up to and including buffalo. Generally, snares are placed along game paths or next to watering holes. Snares are also set for predators around temporary camps in the bush. Hunters then check the snare every day or so. If an animal is still alive, it is despatched with an axe or a machete.[6]

Almost all hunting and snaring is illegal. No one interviewed for this study, for instance, had purchased a hunting licence for any of the animals taken.[7] The purchase of a basic game licence and an area hunting permit requires travel to the distant administrative centre, costs tens of thousands of kwacha and must be purchased for specific species, if those animals are still available on the national quota allocated to an area.

Most hunting forays take place between August and December. At this time of the year, most game congregates along rivers in the Luangwa Valley and is easy located. Once the rains begin and more dispersed sources of water and grazing become available, animals spread out and are more difficult to find or snare. Moreover, by January, flooded plains, muddy paths and swollen rivers make it difficult and dangerous to travel to and from the valley, meat is more difficult to dry, and the humidity and rains may spoil meat being carried back to the plateau. During August to December, full-time hunters make at least one expedition a month while a more ambitious hunter may make two or three trips in a month. In the rains, some hunters continue to make occasional trips to the valley and others continue to hunt in areas near to their villages. The majority of game meat, however, is taken in the Luangwa Valley during the dry season when the animals and terrain are most accessible.

Hunting expeditions tend to be organized by one or two experienced hunters. They purchase the necessary supplies, organize the carriers and decide on a rough itinerary. Once the logistics are in place, the hunters and carriers gather at a designated place and set off for the Muchinga escarpment. It takes approximately 2–3 days of walking to reach the banks of the tributaries to the Luangwa River, where most hunting occurs. Hunters are generally opportunistic, but tend to focus their energy and firepower on the larger animals such as buffalos, elephants and hippos.

Hunters typically continue to search and kill game until they run out of ammunition or they have killed enough meat for each carrier to have a full load – usually at least 10 pieces of meat or 30–40 kg per carrier. Most hunts last about a week (the hunts on which this study is based lasted from 5 to 13 days). As game is relatively plentiful in the Luangwa Valley, expedition members

Table 6.1 **Number of hunters, carriers and wildlife off-takes on separate forays in Luangwa Valley 2002–2003.**

Three hunters, 23 carriers: two elephants and one warthog
Two hunters, 12 carriers: two elephants and one warthog
One hunter, three carriers: one roan, one zebra, one lion
Two hunters, 14 carriers: one hippo and two buffalo
One hunter, 16 carriers: one hippo and one warthog
Two hunters, 18 carriers: one elephant, one eland, two warthogs, one hippo, one duiker
One hunter, three carriers: five duikers
One hunter, four carriers: one duiker
Three hunters, 28 carriers: one duiker, one aardvark, one hartebeest, two elephants, one lion
Two hunters, eight carriers: one warthog
Three hunters, 28 carriers: one warthog, two buffalo, two hippo
Two hunters, eight carriers: two buffalo, one zebra
Two hunters, 12 carriers: one reedbuck, three buffalo, one eland
One hunter, four carriers: two hippos
Two hunters, six carriers: one warthog, one buffalo
Two hunters, seven carriers: one hippo
Two hunters, 20 carriers: two elephants
One hunter, four carriers: one impala, one zebra

seldom return empty-handed unless they have encountered game scouts and had to flee leaving their loads of meat and supplies behind. All of the hunts recounted for this study shot at least one animal and almost all returned with full loads. The scale and quarry of these hunts is illustrated in Table 6.1.

Once an animal is killed, it is generally butchered on the spot, salted, dried and cured over a fire. The carcasses of larger animals (elephants, buffalos, hippos, etc.) are cut into strips and pieces of smaller animals are tied into bundles. Meat is divided between hunters and carriers with the hunter selecting the largest and best-quality pieces for himself, as well as special portions such as the tongue, heart and brisket. Each carrier gets to keep the meat they transport to the plateau, with the exception of the piece(s) that he carries for the hunter.

The game meat trade

I used to go to Mpika four or five times per month from 1995 to 2002. I used to spend about K800,000 per trip, transport inclusive, and I used to get a total of about K2.5 million for loans (debts) or K1.8–2 million on a cash basis. Our clientele are mainly well-to-do people in low density areas in towns – like, in my case, for cash I go to Ndola, but on credit I go to Mufulira and my clients are mainly miners. Some pieces bought at K20,000 in the village sell for K75,000–85,000 there. We usually cut the strips of meat into three pieces of about K25,000–30,000. A limb of a duiker in town now is about K30,000–45,000.   Woman, 53 years old, from Mufulira, 7th-grade education

Hunting is driven by demand for game meat in urban areas as well as the lack of alternative livelihoods in rural areas. Consumers in Lusaka, the Copperbelt and towns throughout Zambia have a strong preference for game meat, and tens of thousands of kilograms of illegally harvested wildlife is consumed each year in urban Zambia. In Lusaka, for instance, it is estimated that the meat from approximately 33,800 impala and 3,000 cape buffalo is eaten each year. In the Copperbelt it is likely that nearly 30,000 impala and 2,600 cape buffalo are consumed annually (Barnett, 2000).[8] In addition, the plateau villages west of the two Luangwa Valley national parks offer strategic access to distant markets via the railway line and a major arterial paved highway.

The demand for game meat in urban Zambia is driven by preference, not cost. Urban Zambians like the taste of game meat and are willing to pay a premium for it. Game meat in urban Zambia is up to 43% more expensive than beef, chicken or goat. This is in contrast to many parts of eastern and southern Africa, where demand for game meat is cost driven. In Kenya, Tanzania, Zimbabwe and Botswana, for instance, game meat is significantly cheaper than its domestic alternatives (Barnett, 2000).

Game meat purchased in the study site is consumed in the Copperbelt, Lusaka, Mpika, Kasama and Kapiri Mposhi. Several dozen locals have become game meat traders and travel periodically from their village homes to Zambia's urban markets. Other locals (particularly those who serve as carriers on hunts) sell game meat on the side of the road. Most large-scale game meat traders, however, are from outside the area. Unlike hunting and carrying, which are exclusively male activities, both men and women are active in the meat trade. The seven women traders interviewed claimed to be single, were the sole support for their children and some relatives, and were attracted to the meat trade by its profitability. Most of them, and the three men interviewed, had some secondary schooling and were the most educated of those interviewed. They realized the risks their enterprise entailed, recognized the high costs of transport and transactions of bribes, yet made trips to purchase meat each month. As the woman whose words began the epigraph to this section concluded her interview:

I expect to make about K2.5–3 million from my sales this journey. I have three children in secondary and two in primary school. They need school necessities. Risk is there, but how to make ends meet is very difficult. Legal business does not make much profit.

Trade in game meat is brisk from September to January, when local hunters are most active. At this time of year, several dozen outside traders visit the study site each week. Most of these traders operate at a small or medium scale; they make 2–5 trips each month and purchase between 40 and 150 kg of meat on each occasion.[9]

Traders purchase meat by size, not by weight. From September 2002 to February 2003 the price for a bundle of dried meat (weighing around 3–4 kg) ranged between K11,000 and K25,000 (US$2.45–5.50). The type of game meat makes little difference to the sale price: elephant, hippo, roan and buffalo meat all generally sell for the same amount. Prices are generally fixed at a particular time, but some bargaining occurs – particularly for the purchase of larger quantities of meat for frequent buyers. Price increases with scarcity of meat and in competition between purchasers. In the past 5 years, the price of game meat has doubled, in tandem with growing demand and inflation. Some exchange of game meat for commodities occurs, with traders swapping new and used clothing, women's dresses and other scarce commodities for game meat.

Once purchased, the meat is placed in maize meal bags and packed into suitcases or travel bags to disguise the contents from game scouts. Although a few traders own vehicles and use these for transporting game meat, most rely on public transport. The TAZARA (Tanzania–Zambia Railway Authority) local service train is the most commonly used method of travel, but traders also travel by bus and lorry. Often drivers and conductors are offered money or meat in exchange for hiding the game meat and for 'facilitating' its passage through Zambia Wildlife Authority (ZAWA) and police checkposts. The going rate for this service is either K20,000 (US$4.50) or a large piece of game meat. If possible, traders like to transport game meat at night when police and scout checkposts are less likely to be manned and the inspections are less thorough.

Once traders reach Lusaka or another urban destination, they divide the game meat into smaller bundles weighing approximately 1 kg. The meat is then sold discretely door to door, from the seller's house to regular customers, or in the markets. Most dried game meat sells for the same price regardless of the species from which it is derived.[10] At the time of research, the price of cut sections from previously dried game meat strips was between K25,000 and K30,000 (US$5.60–6.65). Profit margins for game meat traders are high. Barnett (2000), for instance, notes that Luangwa Valley traders earned profit margins of around 25%. For those selling meat in Lusaka or the Copperbelt, the profits are significantly higher, with traders making up to a 300% profit on the sale of game meat.

The costs and benefits of hunting

> You asked about the difference between poachers and those who don't poach? Those who don't poach are poor. They don't have soap and other necessities.   Young man, 23 years old, 6th-grade education

The sale of game meat is one of the few sources of cash income in the study site. Villagers have few formal or informal employment opportunities. In addition to game meat, the other notable sources of household income are the sale of cassava or tomatoes at roadside stands, the sale of maize, manioc and soya beans to commercial brokers and remittances from relatives in urban areas.

Of the successful hunts recounted for this study, hunters sold game meat valued between K80,000 and K470,000 (US$17–104) and averaged around K300,000 (US$66). Carriers' revenues from these hunts ranged from K160,000 to K280,000 (US$35–62) and averaged around K175,000 (US$39). The most successful hunts can generate total revenues of K4–5 million (US$890–1,100).

Hunts are relatively inexpensive to stage. Carriers take little with them besides maize meal, salt (for preserving meat), a knife, and sometimes alcohol. Hunters' costs are higher as they take these items as well as guns, snares, gunpowder, lead 'bullets' and matches. In October–November of 2002, these hunting inputs cost approximately K29,000 (US$6.45) for each hunter and K12,000 (US$2.65) for each carrier. Considering these financial inputs, the profits from a successful hunt are regularly more than 10 times the amount invested by hunters and carriers.

The 26 hunters whose life histories we gathered in 2002–03 staged an average of 15 hunting expeditions in the year preceding the interviews. They reported that three-quarters of their trips were successful in bringing back expected loads of meat to the village. If the average successful hunt generated revenue of K200,000, these hunters annually earned about K2,200,000 (US$490). Our data on carriers are less robust, but, if we assume that most carriers work an average of five hunts a year, and earn K200,000 per trip, they are likely to earn at least K1,000,000 (US$111) annually.

The sale of ivory also provides some hunters with cash. As highlighted above, ivory used to be the impetus behind much of the 'poaching' in the Luangwa Valley, when most of the flesh from elephants killed was left to rot in the valley. These hunters were locally referred to as 'sokola' – after the Bemba term for pulling teeth. In recent years, however, ivory has become more difficult to sell. Buyers for ivory, however, can be found, and hunters almost always transport the tusks to their villages in the hope of selling them without attracting attention of the wildlife officials or police. Informants noted that in 2002 one tusk weighing about 4 kg sold locally for K200,000 (US$44).

The income generated by hunting also brings indirect economic benefits to those villagers who do not hunt, carry or trade game meat. As most serious hunters do little farming, they purchase or barter most of the food staples (maize, sorghum, cassava) from other local farmers. This in turn provides some farmers with cash and animal protein that they otherwise might not have.

In an area in which per capita income hovers around US$120 a year (Mano Consultancy Services Ltd, 2001), the money earned through the sale of game meat contributes significantly to local livelihoods (Table 6.2). Hunters are among the wealthiest locals. The amount of money earned through this trade, however, is transient and not sufficient to transform their standard of living. The study found that almost all money earned from the sale of game meat is spent almost immediately on food or consumables (clothes, soap, etc.). 'Even earlier participants in the more lucrative rhino horn and elephant ivory trades have retained little to show for their efforts' (Marks, 2001).

Hunting and the game meat trade brings more than monetary benefits to its participants. It also brings nutritional and social benefits. Each of the hunters or carrier interviewed kept one or two pieces of meat for household consumption. This meat contributes to household nutrition levels. However, as many hunters grow little of their own food and rely on the purchase of staples such as maize meal to feed their families, it is unclear whether hunting households are more consistently food secure than non-hunting households.

Social status is another important, but less tangible, benefit of hunting. Historically, Bisa men garnered a great deal of status from their prowess as hunters. This continues to be the case, and many of the most successful hunters are held in high esteem; as patrons, these men use their skill as hunters to protect and to nurture their relatives and clients. In the past, distribution of meat from hunts was expansive, with meat given to a range of kin, neighbours

Table 6.2   **The conversion of game meat into money (Zambian kwacha) based upon the average number of meat strips produced from a single carcass: comparison of prices of meat delivered on the Zambian plateau and in towns, 2001–02.**[*]

Species	No.	(A) Meat strips produced by average carcass (range)	(B) Price of strips from single carcass delivered on Zambian plateau = A × K20,000[†]	(C) Price of strips from single carcass delivered in town = (A) × K50,000[‡]
Elephant, large	5	130 (128–135)	K2.6 million	K6.5 million
Elephant, small	4	110 (96–110)	K2.2 million	K5.5 million
Hippo	5	120 (115–129)	K2.4 million	K6.0 million
Buffalo, male	5	33 (30–36)	K660,000	K1.65 million
Buffalo, female	4	25 (20–28)	K500,000	K1.25 million
Eland	3	43 (38–60)	K860,000	K2.15 million
Roan	2	36	K720,000	K1.8 million
Zebra	3	32 (25–40)	K640,000	K1.6 million
Hartebeest	1	20	K400,000	K1.0 million
Warthog	4	11 (10–12)	K220,000	K550,000

[*]Based upon interview data from hunters and sellers.
[†]Sales price on Plateau ranged between K11,000 and 28,000 per strip depending upon condition, sales competition, and demand (see text).
[‡]Sales prices in town were 2.1–3.0 higher than on the plateau (see text).

and, in some circumstances, the headman or chief. The growth of commercial hunting, however, has transformed social and kinship relations, for, in recent years, many hunters have attenuated their kinship and social relations to reduce their distributional and social obligations. Beyond their immediate dependants, the hunters and carriers interviewed distributed meat to their parents, mothers-in-law, sisters and sisters-in-law and sold the rest. Whether or not a hunter contributed meat to these close relatives at any one time depended upon his pressing obligations and outstanding debts to others who were not related to him.

Hunting is an inherently risky occupation, all the more so when it is carried out illegally (Figure 6.1). Of the hunts recounted for this study, several hunters and carriers had close brushes with lions and charging elephants, one lost a carrier to a cobra bite and several were injured when their muzzle-loaders exploded. The greatest risk hunters and carriers face, however, is being caught, jailed or fined for poaching. The spectre of potential encounters with authorities hangs over all hunts. Local hunters periodically encounter scouts in the Luangwa Valley. Of the hunts analysed in this study for instance, three were abandoned to avoid scout patrols. Of the 30 hunters and carriers interviewed in 2001, seven claimed that their guns had been confiscated, four had spent time in prison and six had spent time in prison in addition to having their weapons confiscated. All of the 26 regular hunters whose life histories were gathered in 2002–03 had experienced run-ins with scouts during their careers. Although most hunters and carriers avoid capture, a number of locals have been arrested and prosecuted for wildlife offences. In 2002, six men from the field site were arrested by scouts. Each received a prison sentence of between 1 year and 18 months. Several game meat traders, including local residents, have also been arrested. In 2002, two local women served 6-month jail terms for selling game meat on the side of the road. More seriously, there have been several recent cases in which game scouts were reported to have shot and killed local hunters and carriers.

Overall, however, the existence of ZAWA enforcement officers does little to deter locals from hunting. As a 40-year-old man told us, 'I poach as the only way to support my family'. Hunters and carriers acknowledge that they risk arrest and jail, yet the relatively slim chances of being caught are considerably outweighed by the direct economic benefits of their participation in the current game meat trade.

Conclusion

The only way of reserving wildlife in order for them to multiply is to employ us poachers so that we can earn money without killing animals. We should be paid for conserving them and for selling meat. If you people just look at us and enjoy apprehending us, it (your complacency) will not work because

Figure 6.1 **A plateau hunter and his young carriers with loads of game meat captured by game guards in the Luangwa Valley in 1989. The guard pointing the rifle set the scene to demonstrate his effectiveness on the job and his resentment towards poaching by outsiders.**

we are poor people. When you come up with measures for dealing with us (anti-poaching patrols, village raids, jail sentences, etc.), we also look for ways to challenge your measures. Under such conditions, hunting and poaching won't be stopped.   Man, 51 years old, 9th-grade education

There are a lot of young men here doing poaching. It's a dangerous business as once our children are arrested there is no bright future for them. They serve jail sentences for long periods. Those jailed don't have money to bail themselves out for the money they get from poaching doesn't do them any good. They use it for such things as beer. Sometimes, these poachers are shot by game guards and no report is made for fear of further arrests. There are so many boys who have nothing to do in the village but hunt.   Woman, 50 years old, married

In Zambia and elsewhere in sub-Saharan Africa, most anti-poaching strategies are based on both carrots and sticks. On the one hand, they seek to improve the effectiveness of scout patrols, crack down on the transport and sale of game meat, and tighten the enforcement of hunting regulations. On the other hand, they seek to offer local residents in a project area alternative sources of income, some of the proceeds from tourism and legal hunts and a greater sense of 'ownership' over conservation initiatives. While sound in theory, this strategy has often been based on assumptions about, rather than knowledge of, the character and extent of illegal hunting in particular areas and the motivations and needs of hunters or enforcers (Gibson and Marks, 1995). With this disconnect in mind, this study provides a snapshot of unlawful hunting and its significance in one set of communities.

The study shows that illegal hunting and sale of game meat has assumed a central place in the economy of the plateau villages surveyed. Hunting appears as the largest local source of cash income, with nearly a third of local households benefiting directly from the game meat trade. No one in the area is becoming rich from hunting, yet, some hunters are able to earn nearly US$100 in a single expedition – close to the local per capita annual income of US$120. Less successful hunters and carriers do not earn this much, but they are still able to earn enough from game meat to supplement their meagre incomes.

The costs of hunting within the adjacent national parks are relatively negligible when weighed against its benefits – even if it is a short-term perspective. The profit from a successful hunt is 10 times that of the investment involved. A possibility of being arrested, fined or even shot by ZAWA authorities exists. However, locals see the probability of this as relatively low and in the face of few alternative sources of income are willing to take such risks.

The long-term sustainability of the national parks in the Luangwa Valley hinges on reducing illegal hunting originating from elsewhere. This will be a difficult task given the profits available to local hunters, the lack of realistic alternative livelihoods and the relatively low risk of being caught and punished by the authorities. Stemming illegal hunting will require much more than improved enforcement of existing hunting regulations. It will require a reduction in the demand for illegal game meat in urban areas and an increase in the legal supply of such meat. At a local level, it will require, at the very least, the creation of income-generating alternatives that are as profitable and as culturally significant as hunting.

In the short term, however, it will be difficult to create livelihoods that will earn hunters anywhere near the same profits they earn currently through the sale of game meat. Any alternative income-generating strategy (tourism, guided hunts, honey production, caterpillar and orchid harvesting and conservation farming have all been vetted) would have to provide local hunters with at least several hundred dollars of income a year to replace lost income from this trade. Moreover, even if profitable income-generating alternatives are created in the area, there is no guarantee that locals will abandon illegal hunting. Hunters may decide that it is in their best financial interest to engage in the new income-generating scheme whilst continuing to hunt.

Notes

1 In this chapter, we use the term 'game meat' rather than 'bushmeat', as this is the term used by Zambian hunters and consumers.

2 This study would have been impossible without the professional and thorough commitment and documentation of this researcher. In an ideal world, he would be credited as an author of this chapter, but with his agreement we decided to withhold his name to protect him from reprisals.

3 Each hunter was asked to describe his age, education, years of hunting experience, the composition of their hunting party (the numbers of hunters and carriers) and weapons used. Each hunter was then asked for a detailed account of their day-to-day activities during the hunt (where they went, what they saw, what they killed, how the meat was divided, etc.), to recount the distribution and sales of meat and to describe how their hunt incomes were spent. Twenty thousand kwacha (about US$4 dollars) was paid to each informant for participating in this study.

4 Some hunting and snaring still takes place on the plateau and in and around the villages. These hunts, however, tend to be relatively small-scale affairs in which individuals or small groups pursue bushpigs or small antelopes at night.

5 For the purposes of this study the exchange rate is taken to be K4,500 to the US dollar. This is the average exchange during the period September 2002 to February 2003. All conversions to dollars are rounded to the nearest US$0.05.

6 All of the hunts on which this study is based relied on muzzle-loaders for killing game. Only one of the hunts reported snaring any animals (a roan and a zebra). Snaring is much more common around the villages on the plateau than used during these long forays into the valley. See Chapter 11 for an assessment of the importance of snaring in the Luangwa Valley.

7 In 2001, we were shown a game licence from the previous year for the following itemized costs: one hartebeest (K75,060), one warthog (K50,040), one bushpig (K10,080), basic licence fee (K100,000) and general receipt (K5,000). For this amount (K240,260 or US$69), the purchaser had to find and kill these species within a 3-month time frame. A woman meat purchaser informed us that she had begun her trade attempting to be legal. She purchased a licence for a hartebeest (K75,000) plus a general receipt and commissioned an ex-game scout, who in 40 days failed to kill one for her. Along with others, her experience convinced her that she was wasting both her time and her money.

8 Statistics for estimated consumption for other species such as hippo or elephant are not available.

9 The lack of large-scale traders could be a product of the stealth involved in smuggling the game meat through police and game scout checkposts. A few bags of game meat are easier to conceal and less costly to lose than an entire lorryload.

10 Our information on the relative prices of different game meat contrasts to that of the TRAFFIC report (Barnett, 2000). This report notes that prices paid for dried game meat from larger animals such as cape buffalo are much higher than for smaller animals such as duiker. This is an issue that requires further research.

PART 2

Institutional Contexts

E. J. Milner-Gulland

Institutions are defined broadly as organizations or mechanisms of social structure that govern human behaviour. Examples of formal institutions involved in the bushmeat sector include national and international legislation governing wildlife use and the wildlife departments implementing this legislation. Informal institutions include traditional rights to hunt and taboos on consumption, as well as commodity chains supplying meat from the forest to markets. The bushmeat trade is largely within the informal sector of countries' economies, and as such is predominantly governed by local and informal institutions as a livelihood activity. However, as a cause for conservation concern, it is an issue for the formal institutions of national government, and is the focus of attention from external institutions, particularly conservation non-governmental organizations (NGOs).

This positioning within both the formal and informal sectors of society is one reason why the institutional context of bushmeat hunting is both complex to disentangle and one of the fundamental drivers of the outcome of interventions. The institutions governing the bushmeat system are multiscale and vary between actors in the bushmeat chain. Impacts of institutions at one scale may be felt at other scales. Thus, the hunter's incentives are shaped by his rights to enter and hunt in a particular forest; his opportunities elsewhere, which depend on his land tenure security; his access to markets, relationship with traders and ability to share the benefits of trade; and the rules which cover hunting and sale of bushmeat, which may come from national or local government, from traditional authorities or taboos. The identity and influence of hunters and traders may be altered by changes in political or social power. This was nicely illustrated by de Merode and Cowlishaw (2006), who showed how commodity chains selling bushmeat from the Garamba National Park, DRC, varied between times of peace and war as power and patronage shifted from female wholesalers to soldiers and back again.

Conservation interventions shift the balance of power and can distort existing institutions or create new ones, the stability of which is a key determinant of the long-term sustainability of any intervention. International organizations can act to change institutions at the national level (e.g. by building capacity in government departments, applying pressure or targeting aid), or at the local level through focused intervention at particular sites. They can choose to work with government, civil society, industry or traditional authorities.

In this part, the authors cast light on different aspects of the institutional structure of bushmeat hunting, and use case studies to illustrate some of the key issues for success or failure of interventions. They also highlight some important areas where we do not as yet have the answers as to how to proceed, and some areas where we have hardly even begun to debate the issues.

In Chapter 7, David Brown gives us a provocative view of conservation policy in relation to the bushmeat issue. His starting point is the gulf between international and national perspectives on the bushmeat trade. The former is overly focused, he argues, on the conservation dimensions – essentially from a negative and repressive perspective – which is quite out of keeping with the positive ways in which bushmeat is often viewed from a national perspective, particularly by the people whose lives are most affected by conservation activities. Some important and uncomfortable questions are raised about the ways in which policy is dominated by external interests whose financial muscle lends them perhaps unwarranted influence on local and national governments. He argues strongly for a shift in perspective towards the national level, with international development assistance providing a major vehicle for this reorientation to come about. However, a comparison with the influence wielded by outside organizations in other sectors may suggest the need to also look at the forces acting beyond the sector – the influence of conservation NGOs, for example, is unlikely ever to eclipse that of international business and financial institutions.

In Chapter 8, Christopher Vaughan and Andrew Long underline the importance of the local perspective, giving us insights from Namibia's experience with community-based natural resource management (CBNRM). This is an approach to conservation that has gained wide acceptability over the last decade, emphasizing the importance of local stakeholders in conservation. The principle is that local users of natural resources, who stand to lose the most from either exclusion or resource depletion, should be given rights over their resources. This rights-based approach is often driven in practice by external agendas, and local involvement is limited to participation rather than control. The authors give a clear exposition of the complexities of implementing CBNRM on the ground. In Namibia, the formal institution of the state provides a framework for limiting but also legitimizing local control of resources, an approach that seems to provide a good compromise that allows effective systems of governance to emerge. The authors make a particularly important point in highlighting the individual-level benefits and costs to people within communities participating in CBNRM. It is easy to consider community benefits as a whole, and lose sight of the fact that in any intervention, however successful overall, there will always be winners and losers. These include people who are excluded because they do not wish to participate in projects, and those whose activities are curtailed by the limitations that the project imposes. It is critical to project success that the viewpoints of these people are considered, otherwise they are likely to undermine the changes. It is also crucial to realize that, in order to obtain public support, control of resource use has to go hand-in-hand with the benefits of this control. Often benefits are only obtained after a time lag, and this means that projects must be initiated based on local trust and goodwill, rather than on clear improvements in livelihoods – which can be a difficult thing to achieve.

The work that the Wildlife Conservation Society (WCS) has been carrying out in the concessions of the timber company Congolaise Industrielle du Bois (CIB) in the Republic of Congo is justifiably famous. In this project, WCS has worked with CIB to improve the sustainability of bushmeat hunting within their concessions, through a broad range of activities including provision of alternative foodstuffs, strict prohibition of certain activities (e.g. snaring, transport of bushmeat out of the concession on company vehicles, hunting of protected species) and formalization of hunting for local consumption. Chapter 9 highlights that, as is the case in Namibia, local participation occurred only within an imposed framework. In this case, however, local control is necessarily more limited than in Namibia, because the project is within a timber concession.

The institutional issues in this case study are particularly difficult, because although there is government support for the intervention, the main implementation is carried out by a multinational corporation, at the instigation of an international NGO. The question of to whom these actors are accountable for their actions is a stark one. On the other hand, in an area such as this, when government is weak, most governance and infrastructure is provided de facto by industries working in the area. In this case, industry is the obvious partner for conservation NGOs to work with, even if ultimate accountability may be to shareholders and supporters in the West rather than to national government or to local people.

In Chapter 10, Andrew Hurst illustrates key institutional issues for successful intervention in the bushmeat system using three case studies. He raises important issues for practical implementation, including the complexities and subtleties of the state as an institution, for example when competing ministries have different levels of power and influence. He highlights the importance of international NGOs realizing that they themselves are part of the institutional context, whether it be lobbying at the ministerial level, or at the microscale level of employing local people and buying food for their personnel.

NGOs are often placed in a very difficult position when intervening in the bushmeat system, because the areas where they work may have dysfunctional institutional structures, unable to provide basic services to the population let alone create and enforce effective and just rules for the management of natural resources. In these situations, the intervention of an international NGO may provide a 'governance envelope' within which good governance is possible. Once the NGO withdraws, the underlying problems of the country reassert themselves. Building functioning institutional structures is slow and expensive, and it is dubious whether it is possible to do for a single subsector of the economy in isolation from all others. This means that, if conservation gains are to be maintained, the NGO cannot withdraw in the short to medium term, and must broaden its remit to include the wider sectors within which bushmeat is embedded. The issues of accountability to the local community and the dangers of being seen as patrons rather than facilitators are stronger the longer the NGO remains, and the broader its sectoral interests.

The issue of sectoral divisions is an important one, which is raised in David Brown's chapter. The trend in international aid is towards government capacity-building, responding to in-country priorities and following poverty reduction strategy plans. As discussed by Neil Bird and Chris Dickson (Chapter 13), this trend is a fundamental challenge to the project-based approach so prevalent in conservation interventions, including for bushmeat. The bushmeat problem is cross-sectoral, involving livelihoods of producers and traders, food supply to consumers and wildlife conservation, and yet intervention has virtually all been from the conservation and wildlife management sector. Unless there is a move towards integrating wildlife conservation into the broader institutional context, not only will progress not be made to strengthen capacity for good governance, but conservation will be increasingly marginalized in the international aid agenda. On the positive side, because bushmeat is so cross-sectoral, and an issue which can attract international interest and finance, it could be an ideal system on which to test cross-sectoral approaches to conservation issues, and convince policy-makers of the importance of true integration of the environment into the agendas of all sectors of government.

We are building an increasingly strong understanding of the drivers of the bushmeat trade, based on the scientific study of hunters, traders and consumers. But there is much less understanding of how the institutional context affects the outcome of interventions, and how interventions themselves feed back into the institutional context of bushmeat exploitation. In this part, and elsewhere in the book, we consider these issues.

7

Is the Best the Enemy of the Good? Institutional and Livelihoods Perspectives on Bushmeat Harvesting and Trade – Some Issues and Challenges

David Brown

Introduction

This chapter is concerned with the role that the institutions of development assistance might play in improving the management of the bushmeat trade. It focuses on the question: Why should bushmeat be of concern to international development policy? The suggested answer is threefold. First, the safety net functions of bushmeat and similar forest products are crucial to the livelihoods of the poor in the humid tropics, particularly in areas with little immediate prospect of transformation out of poverty. Securing these social safety nets is a major challenge, particularly when (as is currently the case) international assistance efforts are focused on social transformation. Second, progress in the management of internationally high-profile and emotive resources such as wildlife may leverage broader benefits in terms of good governance. And, third, to the extent that existing conservation-orientated approaches do not acknowledge the centrality of the bushmeat trade to livelihoods and welfare, they are unlikely to provide sustainable answers to the 'bushmeat crisis'. It is argued that present conservation policy is excessively negative, and needs to develop the positive tone that would come from a more developmental approach.

This chapter starts by looking at the existing conservation strategies in relation to the livelihoods issue, and then considers how the livelihoods dimensions could figure positively in attempts to address the bushmeat crisis within a more positive institutional framework that development assistance lays out.

What is the bushmeat problem?

There is now extensive evidence that, at least in significant parts of the bushmeat heartlands, the volume of trade is likely to pose a major threat to animal conservation. This has damaging implications for the integrity and structure of forest ecosystems. This evidence derives from diverse sources and cumulatively is compelling. What does not necessarily follow, however, is the view that the way to manage this very large and lucrative, if unsustainable, trade is to attempt to ban it altogether. The arguments against such an approach are both welfare related (in terms of livelihood and economic benefits) and practical (the low likelihood of success). Yet this is the strategy that has been most strongly advocated by the conservation lobby both outside and within the main producer states. One of the justifications offered is that the contribution that bushmeat makes to local livelihoods is so small, relative to other activities, as to suggest that curtailing the trade will not impact very negatively on overall human welfare, considered in utilitarian terms. Other data presented in this book point to the significance of wild meat in livelihood strategies across a broad social range, and to the crucial role that this plays in food security. Whichever view is correct, it needs to be remembered that the actors 'at the sharp end' – the forest-dwelling populations – are key decision-makers in this process. Considered from their perspectives, current conservation strategies must seem a nightmarish intrusion into the lives of people with few alternatives, and for whom the penalties for failure to manage risk and secure a livelihood are grim indeed. To argue that it is these people whose behaviour must be the primary focus of change is to demand a very great deal. To support this attitude, as is the dominant approach of much conservation practice, by heavy policing of the poor within exclusion zones – a strategy that offers no incentives to them to invest in managing the resource, and which functions only to further alienate them from any interest in its management – is surely perverse in managerial terms (see Cernea and Schmidt-Soltau, 2003).

I have elsewhere argued (Brown, 2003a) that the starting point in any analysis of the bushmeat trade, and any attempt to better manage it, should be the livelihoods positives not the conservation negatives, and any attempt to improve its management should take their preservation as its fundamental parameter. This is not to downplay the problem of overexploitation, but only to argue that the movement requires local champions, like any other force for change. It is all too easy in the present climate of a crisis to ignore the benefits that bushmeat brings to those at the margins of the modern economy. The virtues of the bushmeat trade include low barriers to entry and high social inclusivity. Unregulated and decentralized in trade, a fair proportion of the value of the product is retained by the primary producer (the hunter). Unlike domestic animal husbandry, the labour inputs which it requires are discontinuous and easily reconciled with the agricultural cycle. Bushmeat is the product of a system of farm/forest management that collectively offers high returns from a range of activities. For the risk-averse small farmers to whom labour is the major constraint, all this has much to commend it. The trade is likewise low risk and flexible, with minimal capital costs, and thus particularly attractive to the poor. Extractive technology is generally low level and accessible. Gender aspects are also positive. In most situations, it is men who do the hunting but women who take charge of all the downstream processing and commerce, to the point of sale in the scores of 'chop bars' and restaurants that are a familiar feature of the urban scene in the south. Bushmeat has excellent storage qualities, in a manner compatible with the storage of agricultural produce. And easily transportable and with an exceptionally high value/weight ratio, bushmeat fits in with the realities of rural life in the tropics in other respects, particularly for the poor. The poor may benefit less than the rich but is this not always the case, and a feature of inequality not of the bushmeat trade?

These benefits are of vital importance in understanding why bushmeat is viewed positively by most strata in the producer societies, and why education of hunters is unlikely to have many positive effects, particularly when it is founded on presumptions of low awareness, rather than as part of a package of measures to help hunters better manage the trade. To adopt the position that these positives must be the starting point in policy development is not to deny the possibility of an impending or actual crisis, but only to shift the focus away from the conservation targets towards understanding the motivations of, and incentives for, the human agents who, at the end of the day, are the only feasible means to achieve them.

The need for a change in approach

The crux of the problem is thus not the conservation science but what happens to policy when one set of interests dominates but when multiple interests are at stake. The fact that those interests are largely generated within northern, post-industrial societies, and imposed on sovereign and largely preindustrial southern states, merely compounds the difficulty. The outcome is a situation in which conservation science does not look a very promising perspective from which to engage local interest or support. The best that can be offered is acquiescence, and this is a poor substitute for ownership. A more positive and respectful approach is required, which seeks to portray conservation science not in opposition to local interests but as their complement.

The grounds for focusing on the local dimension, and treating this issue as primarily a matter for national policy, are numerous and strong:

- The harvest is national and local in character, and within the domain of national and local policy and planning.
- The governance problems to which the high and uncontrolled levels of harvest respond are predominantly national in character.
- The trade is largely national; there may be significant regional and intercontinental components, but these are likely to be adjuncts of the national trade, a relatively small proportion of the total, and not the major drivers of extirpation.

- The external drivers (deforestation, unsustainable timber and NTFP harvesting) are national and subnational in character, and localized in the range states.
- The market for bushmeat is highly specific in cultural terms. It is not likely to extend much beyond the traditional cultural communities; outside the main range states, the market is largely confined to emigrants, and the potential for new consumers is slight.
- Given the importance of bushmeat/wildlife in both local livelihoods and biodiversity conservation, any actions to address their consumption would impinge directly on other sectors and concerns, where the nation-state is the primary rule-setting agency.

Even in the area of policy development related to international treaties and conventions, the main opportunities are located at the national level. Although, for example, the Convention on Biological Diversity stresses that the conservation of biodiversity is a common concern of humankind, it recognizes that nations have sovereign rights over their own biological resources, and will need to address the overriding priorities of economic and social development and the eradication of poverty on a national basis.

The policy focus must thus be a national one, and centred on the producer states. However, given that this is a policy arena in which the dominant forces are very largely expatriate, international development assistance would appear to provide the primary vehicle by which this more positive orientation could come about. The remainder of this chapter is concerned with mapping out the potential for bushmeat to figure in development policy, in support of national agendas for change within the producer states.

Linking conservation strategies to development priorities

The immediate practical challenge is to 'raise the game' and bring bushmeat and other forms of wildlife use firmly into the development policy arena.

There are two broad ways in which this might come about.[1] It might figure on account of its position as a key resource in the livelihoods and welfare of the poor, and it might, in addition, be shown to have potential as an entry point for broader institutional change (for example, governance reform). In the former case, there are also two dimensions and possibilities: livelihood benefits that relate to *survival strategies* and *social safety nets*, and those that have the potential to contribute to *growth and transformation in the rural economy*. In the former case, the main concerns would be to reduce transitory poverty and/or prevent poverty from worsening; in the latter case, the interest is in providing a sustained route out of poverty. It should be noted that these two dimensions are not necessarily alternatives. As will be discussed below, they may well be two sides of the same coin.

Bushmeat and livelihoods

In recent years, the theme of rural livelihoods has occupied a central place in applied development. The UK's Department for International Development (DFID) has been a particularly strong advocate of the 'sustainable rural livelihoods approach' (Carney, 1998), although this has been variously interpreted by its various divisions and partners. Ashley and Carney (1999) identify four broad ways in which it has been conceptualized:

- as a *set of principles* of a largely participatory nature (people-centred, responsive, multilevel, conducted in partnership, dynamic and sustainable in all its main dimensions);
- as an *objective* (and possibly a type of project), aiming to improve the sustainability of livelihoods;
- as a *tool* – for example, building on DFID's 'livelihoods framework' [which relates five types of livelihood assets (the 'five forms of capital' – human, social, natural, physical and financial) to livelihood strategies and outcomes and contexts of vulnerability];
- as an *approach to development*, combining elements of some or all of the above.

The bushmeat debate has clearly been influenced by notions of 'livelihoods' in several of these senses, though perhaps more as an overall perspective than in its more theoretically sophisticated sense.

While the empirical base is still limited, there is now a fairly coherent understanding of the ways in which bushmeat and hunting figure in the welfare of rural populations, particularly the poor. For example:

(a) Significant amounts of data have been collected on the ways in which wild meat figures in the rural economy, both as a tradable and as a subsistence good (see, for example, Bowen-Jones *et al.*, 2002; Hoyt, 2003).

(b) The poor are more likely than the wealthy to depend on bushmeat sales, although the poorest of the poor may be dependent largely on charitable transfers (the term 'the poor' is only relative in such cases, however) (de Merode *et al.*, 2003).

(c) There is strong evidence that lack of access to bushmeat can contribute to malnutrition and declining welfare, thus reinforcing concerns about the depletion of the wildlife stock (e.g. Melnick, 1995).

(d) The availability of alternative protein sources – particularly fish and invertebrates – may mitigate the immediate ill-effects of bushmeat decline (this may help to account for the fact that signals about the decline in wildlife resources are not necessarily acted on) (MacKenzie, 2002; cf. Watson and Brashares, 2004).

(e) As with many NTFPs, the most important livelihoods benefits for the poor are likely to relate to timing and compatibility with the types of multiple enterprise that are the bedrock of the peasant economy, and not necessarily the absolute volume of the trade (Arnold and Ruiz Perez, 2001).

(f) The social profile of the hunter is often the young adult male, without major social ties, and without command of the labour force that would be needed to build up agriculture-based enterprise (Solly, 2004).

(g) Certain types of bushmeat trade can achieve sustainability, as judged by the consistency of the supply over a prolonged period of time (Cowlishaw et al., 2004).

None of this literature demands that a moral stance be adopted on the place that bushmeat ought to occupy in the local economy. There are, for example, no suggestions that increasing the overall volume of the trade will increase the welfare of the poor. However, some conclusions can be drawn regarding an assessment of conservation strategies. For example, the evidence does challenge those who would argue that the welfare of the poor can be secured by permitting subsistence use but banning all commercial trade. It also suggests that, in contexts of limited resources for conservation, a discriminating strategy may have positive effects, moving from broad-brush conservation to well-targeted strategies (for example, conceding hunting of vermin species, and concentrating efforts on the species that are seriously at risk). And it does warn that heavy-handed attempts to cut off the local-level trade could well increase the vulnerability of the poor, have socially disruptive effects and, by implication, be rather unlikely to succeed. It also provides an additional sceptical voice to question the likelihood that alternative income-generating schemes will provide an easy way to alleviate pressure on the wildlife resource (see Ellis and Ade Freeman, 2004).

By and large, livelihoods-orientated research would tend not to commend the small farmer level as the proper target of conservation efforts, given the lack of viable livelihood alternatives. The blanket 'criminalization' of the small farmer in much conservation discourse does little to generate sympathy with external interventions, and the strong opposition in some quarters to the commoditization of bushmeat is also a disincentive to policy-orientated research.

Issues of social risk and vulnerability

The approach from livelihoods provides a useful way for the subsector to articulate with development assistance priorities, at least at the level of social protection measures and social safety nets. Bushmeat is likely to figure strongly among these safety nets. The argument is that, in the types of societies under study, publicly funded social protection programmes do not play a very significant role in supporting the poor and helping them better manage their vulnerability, through labour market interventions, social insurance schemes and the like (Holzmann and Jørgensen, 2000).[2] Public policy plays only a limited part in helping the poor to cope with risk in most of the bushmeat range states, particularly in Central Africa. Rather, the main forms of social safety nets available in these societies come from NTFPs such as bushmeat, largely accessed informally on what are officially 'public lands'. Thus, any strategy that weakens these natural resource safety nets would be likely to prove profoundly anti-poor.

Asset vulnerability has been a theme of growing importance in the development studies literature (World Bank, 2002). There is extensive evidence that vulnerability increases where livelihoods strategies are undermined. This issue has been relatively little studied in relation to bushmeat (but see Melnick, 1995), though it has been the subject of some fascinating research in the field of plant NTFPs. McSweeney (2003; 2005), for example, has produced a penetrating analysis of shock and natural insurance in indigenous households of the Tawahka Asangni Biosphere Reserve in eastern Honduras. She concludes that households invoke some highly differentiated responses to livelihood shocks, and that the nature and intensity of the misfortune experienced, as well as household attributes (such as human capital and land), strongly condition the degree to which forest resources are employed for such purposes. Overall, she reaffirms the centrality of forest products to social safety nets, and underlines the need for realism on the part of conservationists when seeking to curb their use.

An extension of such reasoning may provide an opportunity to champion the cause of the small peasant producer in relation to the safety net functions of bushmeat, and to use this as a rational basis to assert their precedence over non-community interests. This would counter the worrying tendency in the present trajectory of international aid for social safety nets to be discounted in favour of economically transformative activities. But of course the opportunity to insert bushmeat into this debate is available only if the principle of consumptive use is accepted, and if there is a willingness to shift the thinking away from simplistic oppositions of subsistence use (good) and commerce (evil) towards a more subtle understanding of livelihoods and local economy.

There is, then, significant evidence of the importance of bushmeat in social safety nets and their role in reducing vulnerability. This is an important dimension of poverty that needs to be taken into account in poverty reduction strategies. However, donor interest in bushmeat would undoubtedly be much increased if it could be shown that the trade could contribute not just to social safety nets but also to poverty eradication on a substantial scale. It is to arguments around this theme that we now turn our attention.

Potential for poverty eradication?

The role that the bushmeat trade might play in poverty eradication is an under-researched theme in the literature, although there are good reasons to doubt it as a general proposition. A range of factors, including the rustic nature of the commodity, the character and volumes of its trade, and the availability of alternatives, are of relevance here. Paradoxically, it may be partly because of the strengths of bushmeat as a livelihoods asset (low thresholds of entry, leading to broad participation and low profit margins) that it is unlikely to figure strongly in rural transformation. There would also seem to be few opportunities for value to be added in processing or through technical sophistication or increased investments of labour, for this is a good that derives its value from its rustic qualities, and hence excessive processing would destroy its character and dislodge its markets (in this respect, bushmeat may differ from, say, artisanal woodworking). Its transformative potential is likely to be particularly limited where the trade is treated de facto as illegal, and thus has to be pursued underground. The type of capital investment that such illicit trade attracts is unlikely to be concerned with adding local value, and is much more likely in the longer term to adopt an 'asset-stripping' strategy, exhausting the potential of the sector before moving on to reinvest the surplus elsewhere, perhaps with similar asset-stripping intent. Its stigmatization internationally also limits the potential for export-orientated processing and value added, and for the high-value component of a segmented trade to act as the motor for broader transformation.

Viewed from the perspective of volume, bushmeat also offers an unencouraging prospect. Even if the projections of sustainable off-take are overcautious, they are often so far below existing off-take levels as to make it most unlikely that sufficient capital could be generated from the sector to sustain long-term economic change. One of the greatest challenges in the present situation is that any attempt to manage the trade would almost certainly require a substantial diminution in volume of sales, at least in the high-profile cases (charismatics, as opposed to vermin species). This might be partially compensated by price rises, though when the interests of all players are taken into account there is likely to be little prospect of a 'win–win' scenario.

There are some counterarguments, though the evidence base is narrow. For example, in some instances at least, bushmeat does not appear to behave economically as is suggested above, and as would be predicted from first principles. It is usually argued that low barriers to entry to an enterprise push down profit margins, and this diminishes the potential for capital accumulation. Hunting is archetypically an enterprise of this type and, yet, such information as is available suggests that profit margins can sometimes be unexpectedly high. Hunters were found to capture 74% of the final sales price in Takoradi chop bars, for instance, a figure that is similar to those found by other researchers elsewhere in Ghana. Ntiamoa-Baidu (1998) found that hunter income was similar to that of a graduate entering the wildlife service, and 3.5 times the government minimum wage . This compares with Asibey's findings in 1977–78 that hunters earned similar salaries to civil servants, or 8.6 times the earnings of government labourers (see Mendelson et al., 2003).

These unexpected findings may reflect the fact that bushmeat tends to behave as a superior good, particularly in societies in Africa (less so South America), where consumers with purchasing power retain strong affiliations with rural communities (de Merode *et al.*, 2003).

If the principle of sustainability is relaxed as a fundamental principle of intervention, then some interesting questions arise, albeit not without courting controversy. There is a view, for example, that high off-take levels, even risking local extinctions, may be tolerable within a context of broader social change, as long as enough representative biodiversity is conserved. 'Enough' here would imply adequate conservation at both species level and at the level of the genetic stock. An illustration would be the way in which products of the hunt have helped to underwrite the costs of agricultural development in tropical and/or industrializing societies. The expansion of cocoa cultivation in Ghana is an interesting example. In such instances, even a decapitalizing stock could play a part in economic growth and structural change of long-term benefit to the poor (e.g. Asibey, 1977; *see also* Mendelson *et al.*, 2003; but cf. Oates, 1998). This might be regarded as an acceptable price if the net social benefits were positive (which would be particularly the case if the structural changes within society eventually led to the emergence of a pro-forest conservation movement with wide public support).[3]

But even discounting these uncertainties, it seems most unlikely that a significant future justification for donor involvement with bushmeat is going to come from its poverty eradication potential on any major scale, sustainable or not. Its prime livelihoods value looks to lie in its social safety net functions. These functions need to be seen not only as a defensive measure in failing economies, but also as an essential safeguard in expanding ones. It may be precisely in those economies that are undergoing a drive to economic growth, often with development assistance support, that safety nets play their most vital role. This reinforces the view that conservation policy-makers needs to think more, and more positively, about how to secure the requisite core of biodiversity to allow for the time when economic growth lifts the population out of its heavy dependence on extraction of natural resources. Paradoxically, it is this transition that transforms public attitudes to wildlife in the ways that is conducive to modern conservation policy.

Bushmeat and governance

The role that bushmeat might play as an entry point for governance reform could provide another, and complementary, justification for donor interest. Governance is an area in which investment in bushmeat management could 'punch above its weight', by acting as an entry point for reform with wider ramifications, both within the forest sector and more generally.[4]

To say that the lack of bushmeat management is a reflection of poor governance is merely to recognize the chaotic state of the forest sector in many tropical environments. Bereft of their resource rights through appropriation by successive colonial and post-colonial regimes, rural dwellers have had no incentive to manage their (quasi) common property resources sustainably, particularly those resources whose fugitive qualities (in the sense of resources that are not owned until the point of capture) substantially increase the costs and risks of husbandry. It is arguable, moreover, that recent conservation strategies have done little, if anything, to improve the governance environment. Indeed, they may well have worsened it. Drawing their inspiration from, and building on, the disenfranchisement pursued under successive political regimes, conservation institutions have reinforced the inequitable pattern of control over resources (to the benefit of the timber industry, *compradores* in the state and themselves), further diminishing the incentives for forest dwellers to treat wildlife as something they hold in trust. Repressive conservation strategies have also increased the power of state agents to extract rents from villagers, reinforcing their impoverishment with no discernible positive effects on the condition of the resource.

Bushmeat and other products of the hunt nowadays tend to feature among those goods conceded by range state governments, as part of a tacit agreement that separates 'traditional' products for domestic consumption and the generation of lower-level public sector rents from 'modern', industrial commodities that enter into the circuits of national revenue generation and political patronage, and which the population at large has no right to influence. One of the questions that needs to be posed when evaluating new initiatives relating to bushmeat is whether they challenge or support this marginalization within policy discourse. The favoured current strategy is to encourage timber companies not only to ensure that their own practices do not cause environmental damage (which is to be commended), but also (and this is much more controversial) to restrict public access to land areas over which they have temporary logging rights, and to police these heavily on behalf of the state. Tightening up on large-scale forest enterprises may strengthen 'governance' at one level, but it does not necessarily contribute to 'good governance' in the sense of putting in place new institutional rules that are socially just and locally owned, and likely to be respected and sustained. Where existing rights regimes are so profoundly anti-poor, as is generally the case in Central Africa, then attempts to improve governance by consolidating the hold of capital are most unlikely to empower the poor.

The issue of governance reform thus provides an additional area in which a livelihoods perspective is indicated. Recent research on land and wildlife management in southern Africa reaffirms the view that institutional change at policy level is not enough to prevent resource capture by powerful social forces nor to convince local populations that the changes offered are sufficient to give them long-term incentives to invest in proper management (Conway *et al.*, 2002). The evidence from Latin America, where resource rights are often already established, would lead to the same conclusion. The changes have to be anchored in community and individual rights, as the only effective way to convince users that the benefits to be secured by such management will not be abandoned at some future date (Norton and Moser, 2001). Given the very high discount rates on any form of enterprise in the unstable and low-governance conditions typical of the bushmeat producer states, this is likely to be an essential precondition for effective policy reform. Where improvements are framed largely in a policy framework (not as a rights issue), villagers are understandably very sceptical of the degree to which such improvements will stick. Their doubts concern both governments (unstable, prone to change) and conservation agencies/donors (temporary interventions based on 'compassion without obligation', and driven by uncertain external pressures and constituencies).

The central aim of governance reform must therefore be to increase the sense of long-term ownership and responsibility that come from genuine involvement in decision-making around the management of wildlife resource, and the incentives which derive from participating independently in the benefits. Only when public confidence is established will it be possible to address the subordinate, but demanding, changes of practice (discrimination in species selection etc.) that effective conservation will require. This will not come about by further estranging the local populations from the resource, marginalizing their economic interests and welfare, and invoking their participation merely to validate imposed decisions that are against their interests. It is essentially a question of enhancing citizen rights.

Various strategies can be considered to help bring such reform about. One way to address the rights dimension is to focus attention on the subsectoral level, solely on the transformation of hunting rights. Such a subsectoral approach has much to commend it, though it is likely to involve significant and sustained investment costs, and is not for the faint-hearted. An additional difficulty is political vulnerability. Attempts to advance realistic legislation in the bushmeat producer states have proven very vulnerable to political pressures from those who enjoy the fruits of high-level resource control. Given the multiple forces and sectoral interests which act on governance processes in the typical aid-dependent state, subsectoral initiatives are vulnerable even where they succeed in such environments; indeed, perhaps particularly so – for they create the threat of a good example in a way that is easily contained and marginalized. For this reason, they also risk incoherence with other policy initiatives.

All of these factors would tend to further diminish the confidence of the local-level resource users who would be required to invest their time in wildlife management. A more systematic approach to rights, bringing together hunting rights with land and tree tenure changes, and an overall orientation in civil society to the promotion of citizen rights, may well be preferable. It is in this context that the efforts in countries such as Cameroon to involve communities not only in wildlife management, but also in all aspects of forest management, including exploitation of high-value timber, commend themselves. Their chances of success are enhanced by the economic and political benefits of joint enterprise. It is beyond the scope of this chapter to investigate this issue in detail, though there is a growing literature on it, and quite a lot is now known about its political and economic viability (see, for example, ODI, 2001).

Community management is not without its challenges. For example, the scale required for the management of wildlife resources is likely to be an order superior to that needed for profitable artisanal timber management. Given the low levels of social capital common in high-forest societies, this would require careful coordination, and possibly heavy investments in capacity building at the local and municipal level. Similarly, financial viability is also problematic at the local level, in the context of the decentralization policies which are a favoured instrument of government reform. Ellis and Ade Freeman (2004), for example, note the growing importance of decentralization in public policy, particularly in relation to poverty reduction strategies, but warn against the assumption that this will necessarily benefit the poor. Decentralization creates new bureaucratic entities, which are likely to have a thirst for taxation, but little local revenue-generating capacity, given the imbalances in the national economy. Poverty alleviation and revenue generation tend to be opposed, and this could well create pressures for the overexploitation of forest resources. It might also reinforce the alliance between local government authorities and the extractive timber industry, consolidating the concession framework for timber exploitation and blocking any attempts to decentralize control over natural resources in ways that could create pro-poor and anti-industry precedents. Thus, there is a danger that decentralization, within a concession framework, will weaken local government interest in tenurial reform.

However, both of these challenges are ones that have to be coped with, one way or another. The point at issue is not the risks that any sort of policy change entails, but rather the need for politically influential champions to maintain the pressure for sustained governance reform. In all probability, championing of the community interest would have to come from sources other than the government. Experience in countries such as Cameroon has shown how limited is the potential for internally generated reform within the state, given the strength of its alliance with the logging industry. This is despite the existence of individual proponents of reform. Decentralized government may help to redress the balance, but recent research in Ghana warns of the dangers that local government will merely act as a conduit for external voices to be conveyed to the local level. Where the dominant narratives in the society are elitist, then the heavy resource dependence of the decentralized authorities on the political centre tends merely to reinforce the condescending and repressive tendencies of the *mission civilisatrice* (cf. Amanor and Brown, 2003).

What all this cries out for is further strengthening of the incipient convergence between conservation and development that was noted at the start of this chapter. This convergence needs to be pursued with increased commitment and vigour, in relation to field-level policy just as much as in the rarefied committee rooms and conferences in which livelihoods are observed from afar. This convergence needs also to affect relations between conservation and development NGOs. Continued denigration of the livelihoods strategies of the rural populations who depend on the use of wildlife and other forest resources is hardly amenable to this convergence. Nor is implacable hostility to the prospect of legalizing elements of the trade. Nor are all the other elements of conservation strategy that function only to estrange the resource users from the policy process at a time when their support is most urgently required. This convergence must surely be the first step in bringing such a valuable, but anarchic, system of commerce into some sort of order and under some sort of control.

The conservation movement might do well to apply its considerable resources to support the challenge. An alliance between poverty-focused development assistance and conservation, around the joint enterprise management of the full range of forest resources, would be a formidable combination, and a concrete strategy for much needed governance reform. It could well provide the radical force that is required to begin to move forest governance onto a more coherent and rational footing.

Conclusion

The thrust of this chapter has been to argue that a rapprochement between conservation and development assistance is a fundamental requirement if there is to be any chance of addressing the 'bushmeat problem' in a meaningful way. The argument has come full circle, to reassert the importance of the convergence between the two disciplines within a framework centred on human rights. In such a situation, the demands upon the movement that is focused on the resource (conservation) are undoubtedly greater than that for which bushmeat is but one dimension of a broader problem of governance (development assistance). But given the changing architecture and philosophy of international aid, it has to be wondered if approaches that depend so heavily on external influences and funds will be able to survive unscathed, without making concessions to national interests and priorities. Within an international discourse increasingly centred on the targeting of poverty, continued external preoccupation with animal welfare, to the detriment of that of humans, looks to be untenable.

Notes

1 The arguments outlined here are developed further in Brown and Williams (2003) and Brown (2003b). *See also* Davies (2002).

2 These authors estimate that formal social protection schemes of the types that are common in wealthy societies reach less than 25% of the population on a world scale, and very few in tropical Africa.

3 Arguably, this hasn't yet happened in Ghana, though there is a developing 'green' constituency along similar lines to that in, say, the Philippines or Malaysia.

4 This argument is developed further, in relation to forestry in general, in Brown *et al.* (2002).

8

Bushmeat, Wildlife Management and Good Governance: Rights and Institutional Arrangements in Namibia's Community-based Natural Resources Management Programme

Christopher Vaughan and Andrew Long

Introduction

Access to wildlife and bushmeat is important for providing food, cash and a safety net in times of adversity (Bowen-Jones *et al.*, 2002; Elliott, 2002; Rao and McGowan, 2002; Vaughan *et al.*, 2003a; Long, 2004). It also has cultural, spiritual, economic and political value (Hasler, 1996; Bennett and Robinson, 2000; Hulme and Murphree, 2001; Hinz, 2003; Marks, 1976; 2005). However, traditional local institutions for wildlife management have been weakened by both government and NGO conservation policies, making local wildlife utilization illegal. This process has disenfranchised rural people from wildlife ownership and been compounded by unclear institutional responsibilities and ineffective state agents.

Many biodiversity sensitive areas that suffer high rates of bushmeat utilization are under state protection and ownership, such as government protected forests, proclaimed wildlife areas and national parks. Resident or adjacent local communities are often dependent on access to these areas to harvest resources, yet they often have few or no rights to access or harvest creating situations in which wildlife off-takes are unregulated and local communities have little or no control. These factors combine to promote an institutional vacuum of open access to wildlife resources. Consequently, there are few incentives for local users to legally benefit from or maintain these resources and people thus continue to use wildlife clandestinely. In essence, the bushmeat crisis can best be understood as a crisis in the governance of the ecosystem itself (Brown, 2003b).

Recent studies of the bushmeat trade and community wildlife management (CWM) programmes propose not just economic and livelihood incentives but also approaches based on good governance.[1] Recommendations further support the development of new institutional arrangements and rights-based policy and legislation (Agrawal and Gibson, 2001; Hulme and Murphree,

2001; Roe, 2001; Bennett and Robinson, 2002; Elliott, 2002; Grimble and Laidlaw, 2002; Adams and Mulligan, 2003; Brown, 2003b; Adams, 2004; Long, 2004). The provision of economic benefits is commonly perceived as the major reason for changes in behaviour of local people towards hunting and wildlife management (Hulme and Murphree, 2001; DFID, 2002). This chapter challenges this assumption and explores the local rules and practices that influence hunting and wildlife management at a local level.

This chapter utilizes case study material from Namibia's Community Based Natural Resource Management (CBNRM) programme to explore governance arrangements for livelihoods and wildlife management. We argue that changes in wildlife utilization practices and wildlife increases have primarily been achieved by establishing new community wildlife management institutions, deploying community game guards and increasing legal localized control. These changes have been supported by the provision of new local rights, resources and institutional arrangements. However, alongside these new changes, existing local rules or 'norms' continue to influence wildlife management at a local level. Understanding these existing and new influences is critically important for achieving successful CWM programmes. Within the diversity of local rules there are those considered 'best practices' for apprehending and prosecuting illegal wildlife users (Vaughan *et al.*, 2003a). Legitimizing best practices through state and local community endorsement supports community ownership and management. Financial incentives are considered important, but alone are not enough to change wildlife management behaviour. Alternatively, good governance approaches with enshrined rights provision for local communities are central to sustainable CWM.

Governance and institutional issues in bushmeat and wildlife management

Historically, government forest, land and wildlife policies adopted predominantly 'fortress' and preservationist approaches and were legacies of colonial legislation (Adams, 2004). These policies reflected a belief that the rural poor were the root of the conservation problem and were the greatest threat to wildlife, and thus effectively denied communities[2] their customary rights and limited their access to resources (Agrawal and Gibson, 2001; Hulme and Murphree, 2001; Adams, 2004). National governments and conservation NGOs failed to recognize local indigenous knowledge and management practices. This further denied the integral role that customary wildlife utilization plays in local cultural and economic practices (in the context of southern Africa see, for example, Hasler, 1996; Hinz, 2003; Marks, 1976; 2005).

Community attempts to secure resources and exclude illegal exploitation by outsiders were constrained by the lack of national and locally sanctioned enforcement mechanisms, and the absence of transparent decision-making and accountability by state officials.[3] These factors, in addition to a lack of secure tenure and complex overlapping legislation, created an institutional environment hostile to managing wildlife sustainably.

Conservation agents are often unwilling to repeal their own 'fortress conservation' mentalities.[4] In practice, this is influenced by a lack of detailed understanding of the benefits and rationale for a rights-based approach and the unwillingness to value local practices and expertise. Rural households face short-term needs for food security and financial gain. Without resource rights they are unable to make long-term investments in conservation and development based on economic rationality alone. For wildlife conservation to become integrated with rural development, programmes must maintain biodiversity and also demonstrate livelihood improvements in the short term.

To reduce the bushmeat trade and support sustainable CWM there is an urgent need for alternative conservation and development approaches addressing the livelihood priorities of local communities. These approaches should support clarity on institutional rights, authority and increasing accountability by state and conservation agents. New approaches should be supported by informed policies based on negotiated priorities for all parties, thereby offering an opportunity for collaborative and sustainable solutions to the bushmeat trade.

Community-based natural resources management in Namibia

From 1920 until independence in 1990, the communal areas of Namibia were administered by the South African Apartheid government. The pre-independence government paid little attention to wildlife conservation in black communal areas (B. Jones, 1999; 2003). A dual land tenure system entrenched white land ownership, promoting tenure insecurity for communal area residents. Wildlife numbers declined significantly in communal areas of north-west, north-central and eastern Namibia (B. Jones, 2003). The desert-dwelling elephant (*Loxodonta africana*) and black rhino (*Diceros bicornis*) were particularly affected. Factors leading to declines in wildlife populations included increases in human and livestock populations[5] and accompanying habitat loss, lack of appropriate discriminatory legislation, heavy commercial poaching, periods of drought, periods of political instability and the presence of the South African Defence Force (Vaughan *et al.*, 2003a).

Throughout history, wildlife utilization in Namibia has been important to local communities, providing food, income, status, medicine and cultural value. Poorer households and those with limited alternative livelihood strategies often live on wildlife frontlines and endure recurring human wildlife conflict. For these groups wildlife carries costs but also provides an important food security and occasional income safety net. People can use wildlife illegally and supplement their own resources (including livestock), reserving them for when money or food is needed. Wildlife utilization plays a critical, but often secretive, role in supporting social relations, through meat and revenue distribution and as a recreational activity. Historically, traditional leaders had some powers and informal rules to control local hunters (Hinz, 2003), although these were eroded by colonialism, apartheid and other societal changes. During the colonial and apartheid era hunting was common practice amongst government officials and other influential people. During the South Africa–Namibia and Angolan war, pressure on wildlife intensified, and during the 1980s as many as 120 elephant carcasses were recorded by Garth Owen-Smith (unpublished report, 2002). By 1982 fewer than 70 elephants survived out of a previously estimated population of 300 (Viljoen, 1982).[6]

In 1981 responsibility for nature conservation was transferred from Pretoria to Windhoek and the first black game rangers were employed. The Directorate of Nature Conservation began working with traditional leaders and the community game guard (CGG) programme was established. Local community members were employed, trained and provided with rations in return for reporting poaching. These efforts laid the foundations for the development of a more formal and externally recognized system of CBNRM.

With independence Namibia developed a new constitution; within it, Article 95 is dedicated to environmental management and places people centre stage. New political will created space for progressive policy development that addressed community wildlife management. This culminated in pro-community rights-based wildlife legislation in 1996, which was influenced by similar wildlife management devolution activities across southern Africa, including CAMPFIRE in Zimbabwe, ADMADE in Zambia and successes in Namibia's commercial freehold farms. Rights are based on the establishment of new community organizations called conservancies,[7] with a legal basis for community wildlife management in communal areas.

Policies[8] supporting the 1996 legislation[9] set conditions to be met by communal area residents before government registers a conservancy organization and confers rights. To be registered by government, conservancies must elect a management committee, develop constitutions, demarcate boundaries, define membership and develop management and benefit distribution plans. Upon registration, conservancy organizations have conditional and limited rights. These rights are conditional, in that they must meet government criteria, and limited, in that conservancies do not have full rights to utilize all wildlife species (e.g. not for specially protected species including elephant, buffalo and rhino).

The Nature Conservation Amendment Act of 1996 gives limited rights to groups of people to benefit from their wildlife through conservancy formation. Rights are vested in conservancy committees, enabling conservancies to utilize huntable game in several ways, such as trophy hunting, live game sales and own-use meat hunting. Individual personal use of wildlife is still deemed illegal and all wildlife use must be sanctioned by the conservancy. Conservancies undertake a number of activities, including employing staff, developing management plans, and implementing joint venture tourist agreements and community development projects. Community game guards are employed and then deployed on patrols throughout conservancy areas.

Conservancies derive external formal legitimacy[10] from the enabling government legislation and policy. This legitimacy shapes the extent of legal rights, activities and duties of committees and members. Conservancies are de facto acting on behalf of the state and community and can make legal but limited decisions over wildlife management. New conservancy institutional models differ from traditional resource management systems in communal areas. Traditional systems are based on customary law and social, political, ethnic and geographical criteria. The extent of recognition and adherence to new conservancy rules and regulations versus traditional rules and norms forms the critical nexus for new institutional arrangements that are likely to lead to positive behavioural change.

The Namibian CBNRM programme champions the devolution of rights and decision-making for the utilization of natural resources, with a specific focus on wildlife management. Conservancy organizations are able to make use of new rights, in order to manage and utilize their natural resources. CBNRM programme stakeholders support conservancy development activities (e.g. joint tourism ventures, community meat hunting, live game sales, trophy hunting), in order to develop and distribute the flow of benefits from natural resources (mainly wildlife) utilization. These benefits are then supposed to be distributed to community members to act as conservation incentives and a means to offset the costs of living with wildlife, providing increased incentives for changing behaviours towards better wildlife management and conservation. There is a strong economic rationale within the CBNRM programme and focus on developing revenue and cash distribution for conservancy members. The development and distribution of conservancy benefits is seen by CBNRM programme stakeholders as the main means for changing local behaviours.

At present there are 31 registered conservancies, with 14 in the process of registering (Figure 8.1). The government estimates that within the next 5 years almost the entire communal area of Namibia will be under conservancies (Ministry of Environment and Tourism, MET, senior official, personal communication). Over 28% of all communal land is subject to conservancy management, amounting to over 9% of the land in Namibia (Long *et al.*, 2004).

Wildlife increases on communal lands

The CBNRM programme has supported wildlife increases on communal lands, including desert-dwelling black rhino, elephant and popular bushmeat species such as kudu (*Tragelaphus strepsiceros*), gemsbok (*Oryx gazella*) and

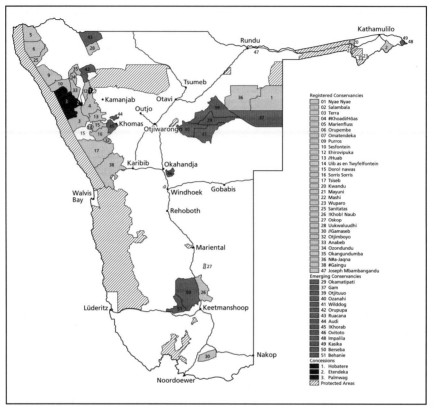

Figure 8.1 **Map of registered and emerging conservancies in Namibia (adapted from Mayes, 2004).**

springbok (*Antidorcas marsupialis*). Since the inception of the programme in the 1990s, there have been no recorded cases of commercial poaching of high-value species such as elephant and rhino, and few cases of illegal commercial hunting and bushmeat trading. Game count data for the last 3 years indicate that there have been 45% and 70% increases in gemsbok and springbok populations, respectively, in the north-west. Table 8.1 illustrates the increases in wildlife numbers.

Many factors have contributed to wildlife increases, including favourable rains, improved governance and organization at the local level, deployment of local game guards, the removal of the army, political stability and proactive external support agencies (Vaughan *et al.*, 2003a). The development of a rights-based CBNRM legislative and policy framework and new institutional arrangements at the local level have been important factors for change (Long, 2004). This change has supported local community wildlife ownership and provided staff, resources and the power to make wildlife management decisions.

Table 8.1   **Wildlife sightings in Kunene, 2001–03 (adapted from Long, 2004).**

Species	2001	2002	2003
Baboon (*Papio cynocephalus*)	144	116	203
Bush duiker (*Silvicapra grimmia*)	13	6	3
Elephant	38	24	44
Gemsbok	1,589	2,616	3,484
Hyena (*Crocuta crocuta*)	1	0	0
Jackal (*Canis mesomelas*)	45	79	60
Klipspringer (*Oreotragus oreotragus*)	4	14	20
Kudu	261	297	241
Ostrich (*Struthio camelus*)	570	659	815
Rhino	1	1	6
Springbok	11,662	14,470	16,733
Steenbok (*Raphicerus campestris*)	54	85	114
Zebra (*Equus hartmannae* and *Equus burchelli*)	1,200	1,274	1,416

Incomes to conservancies from consumptive and non-consumptive tourism (e.g. community harvesting, sales to trophy hunters and live sales) provide financial resources for a few conservancies to develop and manage their own wildlife management institutions and for local distribution. In 2000, the estimated total income for conservancies was approximately N$3.5 million,[11] and in 2003 it was N$14.5 million (Long, 2004). Revenues are being used to cover some conservancy running costs and for distribution to local communities as cash payouts and collective dividends (e.g. school refurbishments). However, the extent to which collective benefits, often to 'urbanized' but remote rural centres, promote changes in individual or household behaviour is as yet unclear and warrants further research (Vaughan *et al.*, 2003b). The majority of rural households have yet to receive any tangible form of conservancy-derived financial benefits that significantly support livelihoods. The distribution of income from revenues, whilst welcome, actually represents a relatively small amount of money and has occurred in a limited number of conservancies. Provision of incentives through collective benefit distribution or individual revenue distribution has yet to be a driving force for changing behaviours towards NRM management (Vaughan *et al.*, 2003b,c; Long, 2004).

Wildlife utilization in Namibia's communal areas

Wildlife utilization provides an important safety net for food security and, to a lesser extent income, particularly for poorer households in communal areas. Conservancy residents were reluctant to give information on wildlife hunting practices, species and numbers for fear of prosecution, making data collection extremely difficult. Data from purposive sample surveys with individual hunters provide an indication of the importance of wildlife at local levels. Surveys carried out in Caprivi and Kunene[12] indicate that in 69% of cases the hunted animals were consumed within the household or shared within people's immediate kin networks as a form of remittance (Vaughan *et al.*, 2004). This compares with only 31% of known cases where animals were sold. The survey data from Caprivi (Mulonga, 2003) indicate that it is primarily small game and game birds that are sold, and that when larger antelope species are sold (following butchering) it is primarily to those with salaries (for example, teachers and government workers).

In the Kunene sample, 84% of the responses stated that hunting was undertaken for the purpose of providing food. The remaining 16% stated that hunting provided food as well as income from the sale of meat. Other data collected in Kunene provide some indication of the way in which particular species are regarded in terms of their food security value. Interviewees from Khoadi//Hôas and Torra conservancies were asked to rank (on a scale of 1–5) the importance of various wildlife species in terms of their value as a resource during times of household shortages. In Torra, springbok were ranked first, followed by spring hares (*Pedetes capensis*), then gemsbok. In Khoadi//Hôas the first ranked species was kudu, followed by gemsbok, then springbok and spring hares jointly third. A range of other species were also ranked but to a lesser extent (including zebra,[13] warthog,[14] cape porcupine,[15] aardvark,[16] jackal, and aardwolf,[17] giraffe,[18] rock hyrax[19] and guinea fowl[20]).

A comprehensive socioeconomic livelihoods and CBNRM survey was conducted across Kunene, Erongo and Caprivi regions (Suich, 2003; Long *et al.*, 2004). The survey covered seven registered conservancies and interviews were conducted in 1,115 households. The households that reported using wildlife and those not using wildlife were compared with reference to key indicators of differentiation and wealth. Indicators included average household income, incidence of receiving pensions and remittances, livestock holding and education.

A total of 619 households were interviewed for the Kunene and Erongo regions. Of these households, 180 stated that they used wildlife, whilst 439 said they were non-wildlife users (Suich, 2003). Based on the chosen indicators of differentiation and wealth, households that reported using wildlife were less wealthy or secure than non-users of wildlife. Wildlife users were less well resourced (had 19.5% fewer livestock) and were less well-off (lower incomes, pensions and remittances) and had slightly less education than the non-user group.

The same analysis of the 496 households sampled in the Caprivi region revealed 127 wildlife using households and 369 non-wildlife using households. Comparing the profiles of the wildlife user group and the non-user group in the Caprivi sample reveals that there are fewer differences and that in most cases differences were slight, with one important exception – livestock holdings. In Caprivi, livestock (cattle) holdings are a well-accepted indicator of wealth (Ashley and Lefranchi, 1997). Twenty-five per cent lower levels of livestock keeping were recorded for the wildlife-using group, strongly indicating that the least well-resourced and least secure of people are wildlife users. These data demonstrate that in both regions wildlife utilization was an activity undertaken predominantly by poorer households.

These findings are corroborated by the following statements from research workshops and interviews with residents of conservancy areas in Kunene.

I only hunt when there is no money, especially when we are desperately in need of food.   Farmer, Khoadi//Hôas conservancy, cited in Vaughan *et al.* (2004)

Those who still hunt are people from poorer households. These are the people with few or no livestock, who have no external support, have no money and are forced to poach due to hunger.   Senior conservancy committee member, Torra Conservancy, cited in Vaughan *et al.* (2003a)

We (residents) are told that illegal hunting is against the law, but I will starve if I don't hunt, therefore I will continue with hunting even if it is illegal.   Young farmer, Khoadi//Hôas Conservancy

The findings above indicate that the wildlife users surveyed in study regions were less secure, or poorer, than the non-users of bushmeat (Vaughan *et al.*, 2004). The extent to which various people utilize wildlife depends on a number of other factors. These factors include the ability to access supportive social networks (which act as a form of social security), geographic proximity to wildlife and conservancy offices and jurisdiction, labour availability, hunting knowledge and skills, extent of poverty and access to alternative livelihood assets (Vaughan *et al.*, 2004). People's preferences for culturally important food and their relationships to conservancy game guards and power networks also affect wildlife utilization practices.

In a further research activity (Vaughan *et al.*, 2004), a purposive sample of hunters in Kunene conservancies were asked if they had changed their hunting or harvesting tactics, with 73% stating that they had changed their tactics since inception of the conservancies. Hunters changed tactics either because they were prevented from hunting or because there was increased local-level control over wildlife or for fear of being caught by game guards.

CGGs and MET staff in Kunene suggested that hunters have adapted strategies of harvesting wildlife to avoid being apprehended by the CGGs, including shifts to more covert hunting strategies such as snaring, only hunting at night and not hunting with dogs. Thus, the control elements of the CBNRM programme clearly restrict people from hunting and can, concomitantly, drastically affect social relations and cause livelihood insecurity.

CBNRM programme stakeholders do not currently acknowledge the effects of conservancy and game guard control activities upon changing the behaviours of the community towards wildlife management. Conversely, CBNRM stakeholders prefer to espouse the belief that people are conserving wildlife and changing their behaviour owing to the provision of economic and other benefits (NACSO, 2001; LIFE, 2002; B. Jones, 2003). However, the CBNRM and conservancies have so far provided limited benefits as incentives and most conservancy residents have yet to receive any financial benefits. In addition, communities and households are now facing the increased risk and costs to their livelihoods of forgoing illegal hunting and wildlife use. Given these factors, the changes in individual and community wildlife management behaviour have most probably arisen not from the provision of financial or other benefits. Changes are rather a direct result of increased localized control by conservancies and the employment and deployment of CGGs restricting local level hunting. This is well summed up by a 23-year-old male farmer in Khoadi/Hoas Conservancy, 'wildlife and us have come a long way together, but now it is our own people watching over us – we are stopped from hunting and are promised benefits but we will have to wait and see'. Conservation gains in wildlife numbers and changes in hunter behaviour have been achieved by transforming local governance arrangements and by establishing mechanisms to control hunting. One critical mechanism for achieving change and new governance arrangements at a local level is the deployment of community game guards.

The role of community game guards

The introduction and establishment of conservancies has fundamentally changed the institutional contexts for local people and wildlife management. A key factor has been the employment and deployment of CGGs. The guards are active within conservancy areas, monitoring wildlife species, numbers and movements, discouraging illegal hunting and apprehending poachers, and are involved in community and trophy hunting and meat distribution. CGGs are the new eyes and ears of the conservancy and are often supplied with uniforms and trained in resource monitoring, apprehension techniques and legal matters.

Currently, CGGs operate within a complex web of different governance authorities (e.g. conservancy, traditional authorities and government ministries) and are accountable to the communities within which they work through conservancy management committees. While the headmen were consulted widely in poaching cases under the original game guards programme (in the 1980s), since the establishment of conservancies, CGGs now report directly to the conservancy committees. Understanding the complex social processes between community members, CGGs and conservancy staff and committee members is critical for understanding how local-level wildlife management practices are influenced and managed.

To explore these processes in more detail, material collected during participatory workshops involving hypothetical role plays illustrate how decisions concerning the apprehension of local hunters were reached (Vaughan *et al.*, 2003a; 2004). The hypothetical role plays involved encounters between CGGs and individuals with different socioeconomic and livelihood circumstances (Box 8.1).

A heated discussion between participants highlighted the diversity of issues affecting decision-making for CGGs. The first issue was that at a community level there are a number of locally accepted (but not legislated) social practices that should be followed. To adhere to local customary practice the guards should first discuss the matter with the headman, then the conservancy office and, if necessary or unavoidable, the MET and, finally, the courts. In the case of the chairman's son (outcome 4 of role play 1 in Box 8.1), the situation was irresolvable at the local level, as the MET officers had discovered the case themselves and were legally obliged to take the matter to the magistrate's court. Had the guard first gone to the headman's house and discussed the issue there may have been other ways to resolve the matter. As the animal was a springbok and the conservancy had a number on their quota they may also have been able to find a solution that would not bring the conservancy into disrepute.

Three comments that were made relating to the scenarios and outcomes in Box 8.1 are noteworthy and bring the dilemmas for game guards into sharp focus.

Going direct to court creates a problem since it means that the community may not give any information in future because there is no chance for the local conservancy to involved in the decisions that are made.   Vitalis Florry, head CGG, Torra Conservancy, cited in Vaughan *et al.* (2003a)

Although the law is the law, we should strive to involve the community in the process. Since the intent of the policy is to give rights to manage wildlife and make decisions, it should also include having a role in these sorts of decisions.   Lucky Kasoana, traditional chief and headman for Warmquelle Conservancy, cited in Vaughan *et al.* (2003a)

The response by MET staff essentially supported the views expressed by CGGs. For example:

The law is a technical problem – conservancy staff and traditional authorities (TA) don't have powers to hear poaching cases. TA has certain powers to act, but with the Nature Conservation Act they have no powers. There is now a

Box 8.1 Role plays of factors influencing poaching apprehension outcomes

The situation

An individual has been apprehended by CGGs and found with springbok meat in the cooking pot (adapted from Vaughan et al., 2004).

Role play 1

The culprit is the son of conservancy chairman, familiar to CGGs from school.

Possible outcomes

1   The man is taken to the Ministry of Environment and Tourism and arrested.
2   A deal is struck between the man and the CGG because the man is the son of an influential person and so may be useful to the CGG in the future.
3   The man is taken to his father, the conservancy chairman, and issued with a warning.
4   The game guard negotiates a bribe and they go drinking together.

Role play 2

The culprit is a poor elderly widow with many dependent children.

Possible outcomes

1   It was agreed ultimately that it would be an MET decision as to whether the woman should be prosecuted or not.
2   The local headman and conservancy chair were involved and the circumstances of the old lady weighed up. She was old, and probably ignorant of the new conservancy objectives. She had many dependent children to feed.
3   The conservancy would arrange for someone to visit the old woman to inform her of the current position with respect to hunting, they would assist the woman, and that if at all possible, they would allocate a springbok from their own quota to her, claiming it as legitimate use.

new wildlife bill. Maybe there will be a change but at the present moment no community members, traditional authorities or conservancy can make civic arrests, there is no [local] forum to finalize the case.   Nahor Howseb, senior MET, Tanger Kunene region, cited in Vaughan *et al.* (2003a)

CGG activities do not take place in a social and political vacuum. CGGs are members of the communities within which they work, and as such are networked into a series of social relations involving kinfolk, traditional leaders, conservancy committee members and other powerful actors. In some instances, CGGs are unlikely to arrest an individual if they have strong social ties with the individual or fear sociopolitical ramifications affecting their social capital, e.g. arresting a relation who may provide the CGG with support for his own livestock farming activities or remittances.

To enable local communities to sustainably manage wildlife, they should be provided the rights and powers to legitimize locally developed and agreed best practices, rules and management practices. If promoted with good governance (increased transparency, accountability, participation and democratic decision-making), it provides great opportunity to develop successful community conservation initiatives.

Conclusion

Managing wildlife use involves managing the governance of social and political relations as well as biological resources and the law. Therefore, achieving community conservation is not simply about technical choices or changes in the law or formal organizations, but is part of wider process involving social change and attempts to redistribute social and political power. The Namibian case study clearly illustrates that governance is at the heart of addressing sustainable wildlife management. Increasing local ownership of resources through devolved management and decision-making provides powerful non-economic incentives for sustainable management. Economic returns to households alone are not enough to promote changes in wildlife management behaviour.

Although national policy and legislation is a fundamental part of sustainable wildlife management, the implementation and development of wildlife management practices is interpreted in locally meaningful ways. Conservation approaches based on governance and rights can significantly contribute to the recovery of wildlife populations. Provision of rights and enabling institutions for wildlife management, with locally accountable and participatory processes and rules, are the foundation for sustainable community conservation. These opportunities should legitimately recognize local forms of wildlife utilization and conservation. It is unclear, however, to what extent conservation and development agencies and national governments are willing to devolve rights to local communities and invest in local institution-building.

To alleviate the bushmeat crisis, new conservation and development solutions are urgently needed. These must meet conservation targets and address the pressing livelihood needs of local communities. This can be achieved only by supporting new institutional arrangements and the devolution of rights to local communities enshrined within good governance and rights-based approaches. Successful conservation and development programmes must implement conservation activities that respect diverse and complex political, ecological and socioeconomic landscapes. The bushmeat crisis is after all a crisis of governance, played out in an ecological environment. Solutions require substantial investment in infrastructure and technical goods. However, for solutions to truly succeed conservation and development programmes must invest in social and human capital, good governance, new institutional and organizational arrangements and the provision of rights for local communities.

Notes

1 The International Union for the Conservation of Nature (IUCN) Commission on Environmental, Economic and Social Policy (CEESP) defines 'governance of natural resources' as the interactions among structures, processes and traditions that determine how power and responsibilities are exercised, how decisions are taken, and how citizens or other stakeholders have their say in the management of natural resources – including biodiversity conservation. It considers good governance 'to include clear direction, effective performance and accountability, and rests on fundamental human values and rights, including fairness, equity, and meaningful engagement and contribution to decision-making' (http://www.iucn.org/themes/wcpa/theme/governance.html).

2 The use of the term 'community' by the authors does not imply any sense of a homogeneous social and political entity. The term community is used here as shorthand, but reflects the inherent social heterogeneity and complexity within rural populations which broadly share the same historical origins.

3 Recent evidence from Cameroon supports this view, and it is clear that when local hunters and leaders are involved in designing and implementing management systems more effective control is possible. In southern Cameroon, local community arrangements to collectively hunt and process meat for sale to labourers working in the logging camps has led to a dramatic reduction in illegal poaching by outsiders (WWF presentation to DFID, 17 February, 2005).

4 This would mean a major shift in conservation organizations' agendas and public image. At this stage it is unlikely to happen as the 'save it' crisis narrative is much stronger than the 'use it' voice.

5 This was a consequence of the policy of establishing 'homelands' for ethnic groups under the apartheid system.

6 Compared with other parts of Africa the densities of wildlife in the arid north-west of Namibia are much lower.

7　A conservancy consists of a group of commercial farms or areas of communal land on which neighbouring land owners or members have pooled resources for the purpose of conserving and using wildlife sustainably. Members practise normal farming activities and operations in combination with wildlife use on a sustainable basis. The main objective is to promote greater sustainable use through cooperation and improved management. Conservancies are operated and managed by members through a conservancy committee (http://www.met.gov.na/programmes/cbnrm/cons_guide. htm).

8　The Wildlife Management, Utilization and Tourism in Communal Areas Policy of 1995 (Government of the Republic of Namibia, 1995).

9　The Nature Conservation Amendment Act 1996 (Government of the Republic of Namibia, 1996).

10　A distinction is made between formal external and informal internal. One is the recognized legal rights provided by the state whilst informal internal rights maybe those imposed by the local community and may not be enshrined in any legislation but are recognized as accepted social behaviour by the broader community. Often informal internal legitimacy is more powerful at shaping outcomes and activities when dealing with secretive issues such as illegal wildlife utilization in an environment of mistrust such as post-apartheid Namibia.

11　At the time of writing US$1 was worth 6.55 Namibian dollars.

12　In Kunene this sample included a total of 37 individuals who were known hunters from Torra and Khoadi//Hôas Conservancies ($n = 13$ and 24 respectively). In Caprivi the sample involved 39 individual informants in Salambala ($n = 15$) and Mayuni ($n = 11$) Conservancies and an area outside of the conservancy boundaries in Linyanti ($n = 13$) who were known hunters and wild food harvesters.

13　Zebra (*Equus burchelli*).

14　Warthog (*Phacochoerus aethiopicus*).

15　Cape porcupine (*Hystrix africaeaustralis*).

16　Aardvark (*Orycteropus afer*).

17　Aardwolf (*Proteles cristatus*).

18　Giraffe (*Giraffa camelopardalis*).

19　Rock hyrax (dassie) (*Procavia capensis*).

20　Helmeted guinea fowl (*Numida meleagris*).

9

Wildlife Management in a Logging Concession in Northern Congo: Can Livelihoods be Maintained Through Sustainable Hunting?

John R. Poulsen, Connie J. Clark and Germain A. Mavah

Livelihoods, sustainability and logging concessions

People living in subsistence economies have hunted wildlife for millennia (Bahuchet, 1993). Patterns of wildlife use are changing, however, with commercialization of the bushmeat trade providing economic incentives for greater harvest of forest animals than before. Where logging takes place, pressures on wildlife populations increase dramatically (Robinson *et al.*, 1999). Immigration of workers for the logging industry increases the human population, leading to a greater demand for bushmeat. Indigenous hunters often shift from traditional hunting methods to more efficient weapons (Wilkie and Carpenter, 1999). Logging roads open the interior of previously inaccessible forest to hunters, giving access to new sources of wildlife. Moreover, logging trucks provide transportation of carcasses to markets, reducing the production costs of the hunter and increasing labour efficiency (Auzel and Wilkie, 2000). And as timber companies facilitate commercial bushmeat trade, the possibility of obtaining income from hunting attracts people to the trade. As a result, traditional systems of resource management often break down and local populations are marginalized from management decisions, especially in situations where logging companies privilege their workers over non-workers.

With 45% of tropical forests allocated to logging companies (Global Forest Watch, 2002), this scenario of intensive commercial hunting, unsustainable harvest of wildlife and breakdown of traditional wildlife management systems probably describes many sites across central Africa. The challenge then, is to develop a wildlife management system that (1) mitigates the negative effects of logging on wildlife, (2) encourages sustainable hunting systems that support local people's livelihoods, (3) reinforces the capacity of forest people to manage wildlife populations and (4) protects wildlife populations, particularly those of vulnerable and endangered species.

In the Republic of Congo, the Project for the Management of Ecosystems in the Periphery of the Nouabalé-Ndoki National Park (PROGEPP in French) has been created to manage wildlife in four forestry concessions (Kabo, Pokola, Loundoungou and Toukoulaka; Figure 9.1). PROGEPP, a partnership of the government of Congo, the Wildlife Conservation Society (WCS) and the Congolaise Industrielle des Bois (CIB), has created a wildlife management system based on four key principles: regulating access to wildlife resources, promoting selective hunting, involving communities in wildlife management and developing alternatives to hunting.

Regulating access to wildlife resources: the who and where of hunting

Whereas access to timber resources is well defined by Congolese forestry code (Law 16-2000), wildlife within forestry concessions throughout Central Africa generally remains a common pool resource with few or no restrictions on access. A wildlife code exists in Congo, but wildlife regulations are rarely enforced. One way to reduce the risk of overexploitation of wildlife is to restrict who has the right to hunt and limit where hunting can occur. In villages such as those found in Central Africa, the level of dependence on wildlife may have created the situation of hunters curbing their own use (Runge, 1981; 1984; Dietz *et al.*, 2000). Village institutions often provide mechanisms to govern resource exploitation. The effectiveness of these mechanisms, however, depends on

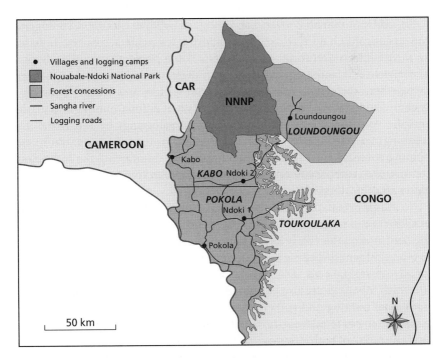

Figure 9.1 **Map of the Nouabalé–Ndoki National Park and the forestry concessions attributed to CIB in the northern Republic of Congo.**

strong village institutions and interpersonal relationships (C. M. Rose, 2000), both of which may fail when large numbers of outsiders immigrate into an area or when hunting becomes a commercial activity rather than a subsistence activity.

In northern Congo, the population density has increased with the immigration of workers for logging. In the five logging villages within the CIB logging concessions, the population rose by 70% from 1999 to 2005 (from 10,122 to 17,979 people). Of the total population (30 villages and 102 camps) of 28,264 in 2005, 45.6% were immigrants from outside of the Sangha region. Immigration has increased the demand for bushmeat, and has also strained traditional systems of natural resource management. Immigrant members of the community rarely possess the same incentives to control their consumption of common pool resources as indigenous people because they do not identify with the land or the community and may not have a long-term interest in the area. These same immigrants are disproportionately influential in the management of wildlife resources because they work for the logging company, and thus have wealth and prestige and are well organized compared with non-workers. For example, CIB employees are organized in unions that lobby the company and government to protect their rights and increase their benefits. In fact, the wildlife and hunting rules adopted by CIB (Box 9.1) incorporate specific benefits (bimonthly controlled hunts and certain alternative activities) for workers as a result of negotiations with the workers' union.

To address potential over harvesting of wildlife in the CIB logging concessions, PROGEPP worked to establish a restricted access system based on the traditional territories of indigenous Bantu populations and traditional movements and land use patterns of the indigenous semi-nomadic Mbenzele living within the region. After meetings with villages and Mbenzele camps, a system of hunting zones was delineated based on traditional hunting boundaries and the seasonal forest forays (hunting and gathering) by Mbenzele. The resulting restricted access system limits immigrant access to the traditional hunting zones of most indigenous communities, reinforcing the indigenous community's authority over these zones and preserving access to nearly the entire area for Mbenzele.

To a large extent, the hunting zones take into account the different cultures of villagers and semi-nomadic (Mbenzele) peoples. Whereas villagers perceive the land around the village as territory belonging to the village and off-limits to others, Mbenzele believe that the entire forest was created by *Komba* (God) for all to share (J. Lewis, 2002). Therefore, by delineating village hunting zones (in which Mbenzele are also permitted to hunt), and at the same time allowing traditional hunting and gathering throughout most of the concessions, the system attempts to address the traditional access systems and perceptions of land use rights of both cultures.

Three types of wildlife use zones have been created: village hunting zones, conservation zones and protected zones (Figure 9.2). Village hunting zones are reserved for the hunters within the adjacent village and are subdivided

Box 9.1   Hunting rules in logging concessions attributed to CIB

In addition to Congolese hunting laws (Law 48/84), the logging company adopted rules for its employees that reinforce and even exceed Congolese hunting laws. Most notably, the restriction on exporting bushmeat from one site to another is not part of Congolese law, although it was enunciated in a policy statement by the Minister of Forestry, Economy and the Environment (MEFE) and therefore supports national-level policy. The rules are intended to maintain biological diversity and protect habitat in the concessions, protect endangered species, ensure sustainability of wild animals and reduce the indirect impacts of logging on the Nouabalé-Ndoki National Park (CIB, 2006). Employees that break company wildlife rules may be penalized, possibly losing part of their pay or even their jobs. Company rules include:

1   The hunting of protected species and the use of snares made of metal or nylon are prohibited.
2   Workers must obtain a hunting permit and licence to carry a firearm.
3   The transport of hunters, firearms and bushmeat in company vehicles is banned.
4   Drivers are responsible for the people and materials transported in their vehicles (and therefore drivers can be penalized if they carry bushmeat or hunters).
5   Drivers must stop at control posts and permit ecoguards to search vehicles.
6   Driving at night without written authorization is prohibited.
7   Protected and conservation zones where hunting is prohibited must be respected.
8   The export of bushmeat outside the zone where it was captured is banned (i.e. only local consumption of wild meat is allowed).

into zones for villagers, residents of logging sites and the controlled hunt (see below). The conservation zone prohibits hunting with firearms, but permits hunting and trapping with traditional weapons and fishing and gathering throughout the year. Protection zones are areas that are of particular importance for the conservation of large mammals (the buffer around the park borders and large forest clearings and mineral licks) and all hunting is off-limits. The conservation and protection zones serve to protect populations of game and key habitat, and presumably as a source of wild animals to replenish wildlife stocks in neighbouring hunting zones (McCullough, 1996; Fa and Peres, 2001).

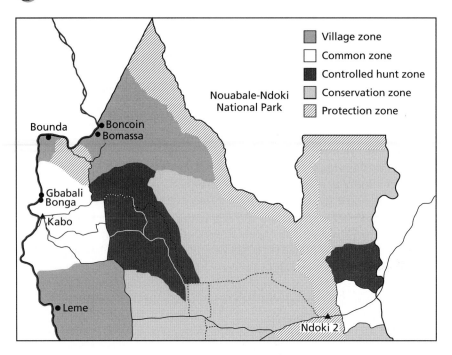

Figure 9.2 **Maps of hunting zones for the (a) Kabo concession and (b) all concessions attributed to CIB. The zonation plan for Kabo has been formally adopted by the government of Congo within the Kabo management plan. It is divided into protected zones (413 km², 14% of the concession), conservation zones (1,154 km², 39% of the concession) and village hunting zones (1,396 km², 47% of the concession). The zonation plans for the Pokola, Toukoulaka and Loundoungou have not yet been formally adopted.**

This zonation system was formalized in the Kabo concession with the adoption of the Kabo management plan by the government in 2006 (CIB, 2006). In the form proposed by the logging company and adopted by the government, the zonation system strayed from its goal of reinforcing the rights of indigenous people in two ways. First, although the village zones were originally intended to provide access only to indigenous people (i.e. the original inhabitants of the village), all residents of villages, including immigrant CIB workers, are permitted to hunt in the zones. Thus, in some areas (specifically around Kabo and Pokola), forest areas that traditionally supported a couple of hundred people are now called upon to feed a few thousand people. Second, controlled hunt zones are reserved for bimonthly controlled hunts organized specifically for CIB workers (despite the fact these workers have the legal right to hunt in the 'village' zones with which they are connected). A group of workers is transported in a company vehicle to the controlled hunt zone, thereby privileging workers over non-workers, who are not allowed to take part in the hunt. Although far from perfect, the zonation system is a first step towards guaranteeing the rights of local people to hunt, while trying to manage for wildlife in the long term.

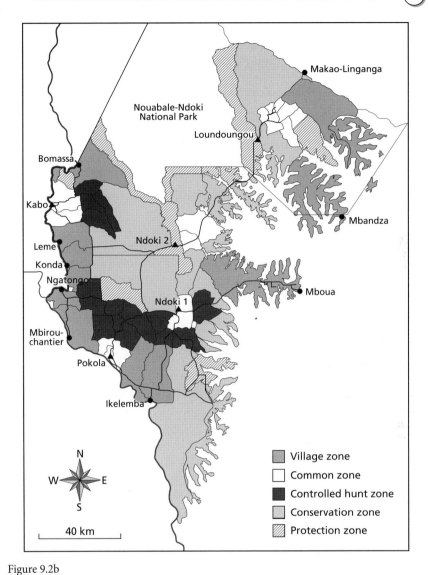

Figure 9.2b

Promoting selective hunting: the how and what of hunting

Whereas the zonation system regulates who can hunt and where they can hunt, selective hunting reduces the risk of overexploitation by regulating which species can be hunted and how many animals can be harvested. The Congolese Wildlife Law clearly defines game species and quotas, although the law is difficult to enforce. The law defines three classes of species. Protected species cannot be hunted at any time. Partially protected species, e.g. forest buffalo (*Syncerus caffer nanus*) and yellow-backed duiker (*Cephalophus sylvicultor*), can be hunted with a large game hunting permit. All other wildlife species can be hunted using non-traditional weapons, within a predefined quota; however, the quota is not enforced throughout Congo. Traditional hunting techniques (crossbows, spears, nets, traps constructed with non-imported materials) are legal with no limit to the number of individuals that can be killed.

To promote selective hunting, PROGEPP ecoguards enforce many Congolese hunting laws, focusing on two principles: (1) no hunting of protected and endangered species; and (2) no hunting with snares – because they are indiscriminate in the prey they trap and kill. With the knowledge that people, particularly Mbenzele, hunt for subsistence, other hunting laws (e.g. non-hunting season) are not strictly enforced. The enforcement of Congolese hunting laws may become stricter over time as people become cognizant of hunting rules and accustomed to the presence of ecoguards. PROGEPP raises awareness of hunting rules through village meetings, dissemination of posters, information campaigns, radio broadcasts and theatre. In addition, WCS has written a book and activity guide on the endangered species of Congo which is currently being taught in schools and Mbenzele communities throughout the CIB logging concessions. Thus, a strategy of combining awareness-raising and enforcement promotes selective hunting.

Involving communities in wildlife management

Too often, restricted access systems (e.g. demarcation of logging concessions, national parks, hunting zones, etc.) are decided by high-level decision-makers without the participation of local communities, leading to their loss of both legal and practical land use rights in the process. When people are removed from the process, compliance tends to be low, and conservation resources must be disproportionately allocated to enforcement. PROGEPP involved local communities in decision-making during the development of hunting zones (see above). Most other decisions involving wildlife management were decided at an institutional level by the government, CIB and WCS. The institutional relationship between the partners is described in Box 9.2 and in Elkan *et al.* (2005).

The goal of PROGEPP is to evolve towards a locally managed solution whereby sufficient incentives and capacity exist for local people and local law enforcement to work towards the sustainable management of wildlife. By incorporating local communities into the development of the restricted access system and by strengthening the authority of indigenous communities to manage resources within their hunting zones, law enforcement by ecoguards could eventually take a back seat to village-based management mechanisms. As a means of reinforcing village institutions and formally incorporating indigenous populations into resource management, PROGEPP has begun to organize resource management committees of villagers and semi-nomadic Mbenzele. Resource management committees offer a conduit for information exchange with local communities and a structure for implicating people in the development of hunting rules and zones. In this way, PROGEPP seeks to empower communities to make decisions that concern wildlife management

Box 9.2   Institutional relationships among PROGEPP partners

Situation pre-PROGEPP

Indigenous Bangombé and Mbendzelé pygmies originally subsisted in the area as semi-nomadic hunter–gatherer societies at low population density ($<0.5/km^2$). With the arrival of logging companies in the late 1960s, permanent settlements were established along the Sangha River, especially near the commercial forestry bases of Pokola and Kabo. Civil war during the 1990s caused even further immigration to the area because Pokola was not influenced by the violence. Hunting camps lined the major logging roads and it was common to see piles of bushmeat both in the Ouesso market and on the aeroplanes to Brazzaville.

Institutional framework and initiatives

In 1998, as logging operations around Nouabalé–Ndoki National Park increased, WCS submitted a zoning proposal to the government of Congo to protect the park from hunting and demographic pressure. The government rejected the plan and asked WCS to negotiate with CIB to develop a wildlife management and protection plan reflecting the government's twin goals of biodiversity conservation and revenue generation through timber exploitation. In 1999, the government of Congo, CIB and WCS signed an agreement forming PROGEPP. As a result, CIB modified its company rules and management to incorporate and promote wildlife conservation. The agreement also resulted in the creation of a protection unit (ecoguards) to enforce Congolese laws and the company wildlife rules (Box 9.1). WCS and the MEFE manage PROGEPP conjointly, administering the project and developing and executing project activities. The project is primarily funded through donations to WCS by private and institutional donors. CIB contributes approximately 15% of total project costs, with the majority of its contribution funding law enforcement (ecoguard salaries, etc.). PROGEPP and CIB meet monthly to discuss technical issues, particularly those that specifically involve the company and its workers (rules to control use of logging roads, controlled hunt and alternative activities, etc.).

Outcome

The post-PROGEPP context is very different from the pre-PROGEPP context. The wildlife management system described in this chapter is solidly in place, formally adopted by the government for the Kabo concession (CIB, 2006), and understood by all partners and the majority of local people. Hunting camps no longer line the logging roads. And, although poaching still takes place, it has been much reduced compared with the pre-project situation. From an institutional perspective, the initiatives taken by the project contributed to the recent FSC certification of the Kabo concession. FSC certification allows CIB to gain a higher price for its wood as a direct result of improved environmental management. At the same time, the company must continue to manage forests and wildlife in a sustainable way to keep its certification. Even though certification criteria do not yet adequately address wildlife issues particular to the forests of the Congo Basin (Bennett, 2000), it is the first step towards passing on the costs of biodiversity protection and wildlife management to industry and customers.

(e.g. developing hunting rotations around villages, reducing harvest of rare species or developing systems to restrict the use of hunting zones by outsiders, if necessary). At the request of communities, PROGEPP researchers could continue to provide technical assistance to committees while ecoguards could reinforce village efforts when local mechanisms to control illegal hunting in their zone fail (e.g. help control 'outsiders' hunting with military weapons). For Mbenzele, who often do not reside in villages, but rather move freely throughout the concessions, management committees will serve as an important tool to improve communication between PROGEPP and the semi-nomadic population. The forest lifestyle and semi-nomadic culture of the Mbenzele have led to a relative lack of formal organization and representation compared with villagers (J. Lewis, 2002). At present, policy decisions (e.g. determining which areas are to be set aside from logging or hunting or where and how CIB workers can hunt) are made by the logging company, the government, the project and elite members of villages. Resource management committees will hopefully ensure that the Mbenzele, like the villagers, can be involved in policy decisions.

Developing alternatives to hunting and wild meat

Through education, consumers frequently change their preferences for goods that provide little to diets (e.g. ivory, gorillas and chimpanzees), but the use of goods that satisfy a basic human need is unlikely to change (Freese, 1998). Given that wild animals are one of the principal sources of animal protein in northern Congo, demand for bushmeat is unlikely to change unless substitutes are available. Because rural people gain food and income from hunting (de Merode *et al.*, 2004), alternatives to hunting and bushmeat must include substitutes for sources of protein and household income.

Box 9.3   Densities of some protected species in logging concessions

Large-mammal surveys conducted in the concessions from 2000 to 2002 suggest that these areas contain densities and abundances of large mammals (elephants, gorillas and chimpanzees) roughly equivalent to those in the adjacent Nouabalé–Ndoki National Park and Dzangha–Ndoki National Park (Poulsen *et al.*, 2005a–c). Estimates of densities of these species in the neighbouring protected areas range from 0.6 to 0.90 elephants/km² (Carroll, 1988; Fay and Agnagna, 1991), from 0.89 to 2.14 gorillas/km² (Carroll, 1988; Fay and Agnagna, 1992; Blom *et al.*, 2001; Morgan *et al.*, 2006) and from 1.21 to 1.93 chimpanzees/km² (Morgan *et al.*, 2006). Chimpanzee densities may be lower than in the Nouabalé–Ndoki National Park as a result of tree extraction and habitat modification during logging (Poulsen *et al.*, 2005a).

Species	Site	Density of individuals per km² [95% CIs]	Abundance of individuals [95% CIs]
Forest elephant*	Kabo	0.74 [0.62, 0.89]	3,578 [2,985, 4,290]
Loxodanta Africana	Pokola	0.83 [0.65, 1.06]	3,950 [3,093, 5,031]
	Toukoulaka	0.16 [0.11, 0.24]	465 [312, 694]
	Loundoungou	0.31 [0.09, 1.05]	1,082 [333, 3,635]
Western lowland gorilla	Kabo	1.36 [1.05, 1.75]	3,605 [2,985, 4,290]
Gorilla gorilla	Pokola	2.15 [1.51, 3.06]	8,013 [5,627, 11,404]
	Toukoulaka	2.25 [1.60, 3.17]	4,627 [3,285, 6,518]
	Loundoungou	0.56 [0.04, 8.23]	1,554 [111, 22,841]
Chimpanzee	Kabo	0.29 [0.24, 0.35]	842 [701, 1,011]
Pan troglodytes	Pokola	0.35 [0.28, 0.43]	1,304 [1,043, 1,602]
	Toukoulaka	0.44 [0.35, 0.57]	913 [713, 1,168]
	Loundoungou	0.30 [0.11, 0.78]	833 [305, 2,165]

*Using elephant dung decay rates based on precipitation patterns in the concessions rather than literature reported decay rates provides less conservative density estimates of 0.23–1.23 elephants/km².

PROGEPP has focused on small-scale extension of techniques to improve vegetable gardening, fishing and animal husbandry practices, as well as experimenting with new animal husbandry practices for the area (guinea pigs, porcupines, rabbits and snails). Fishing supplies are supplied to fishermen at low cost. Vaccination of chickens against bird pests supports the development of small-scale poultry farming. At a larger scale, PROGEPP worked with CIB to import beef and frozen meat to logging villages, including building refrigerator chambers and a butcher shop (Elkan *et al.*, 2005).

Preliminary results of wildlife management

Are current levels of wildlife harvest in the logging concessions sustainable?

Populations of some of the most endangered species in central Africa appear to be doing relatively well throughout the logging concessions (Box 9.3). However, current levels of consumption and trade of legally harvested wildlife species may not be sustainable. Here we examine the sustainability of hunting around logging towns using two different methods: extrapolations of market data to the total off-take of duikers and comparison of harvest rates per hunt over time and between sites. Currently, it appears that the sustainability of hunting around logging villages depends on the site and population pressure.

Duikers are the principal prey species in the logging concessions, accounting for 68% of all prey species annually across four logging villages. The average production rate of duikers across the Congo Basin is 170 kg/ km²/year (Wilkie and Carpenter, 1999), although this exceeds the theoretical estimate of productivity of a tropical forest derived by Robinson and Bennett (2000). Assuming that relatively short-lived animals should not be harvested at a rate exceeding 40% of annual production (Robinson and Redford, 1994), the sustainable harvest of duikers is approximately 68 kg/km²/year. Auzel and Wilkie (2000) reported that 36% of animals hunted in Ndoki 1 were sold in markets. Therefore, using 36% as an estimate of the proportion of captured wild animals that end up in the market, the amount of bushmeat in markets should not exceed 24.5 kg/km²/year.

From 2001 to 2004, an average of 10,829 kg/year of duikers was found in the Kabo market. With a hunting zone of 572 km², duikers appear to be harvested at a sustainable level (18.9 kg/km²/year) around Kabo. If the area that is farther than a day's walk from Kabo is removed from the hunting zone area, then the hunting area is 416 km² and the yearly off-take of duikers is 26.0 kg/km². This second estimate of off-take is probably closer to the reality. From 2001 to 2004, an average of 42,933 kg/year of duikers was recorded in the Pokola market. With a hunting zone of 830 km², 51.7 kg/km² of duikers is harvested each year near Pokola. This level of off-take, which is more than twice the estimate of sustainable off-take, overestimates the actual area in which hunting occurs.

Similar to the market data, data from the controlled hunts around Kabo and Pokola suggest that the forests around Pokola may be overhunted. Several studies report a consistent transition towards smaller-bodied prey species over time as hunting reduces abundances of slower-reproducing, larger-bodied species (Jerozolimski and Peres, 2003; Fa *et al.*, 2005). Optimal foraging theory suggests that only when more valued (larger) species decline in abundance would less valued (smaller) species be added to the prey set (Stephens and Krebs, 1986). An evaluation of hunter prey sets (off-take from controlled hunts) indicates that the percentage of medium-sized duikers (14–24 kg) hunted in Pokola ($\bar{x} = 32\%$, range = 27–47%) forests is half that of Kabo ($\bar{x} = 63\%$, range = 59–70%). By comparison, on average, monkeys, weighing 3–13 kg (Pokola $\bar{x} = 17\%$, range = 7–24%, Kabo $\bar{x} = 12\%$, range = 9–14%), and other small species make up a greater percentage of the species hunted around Pokola than around Kabo. Other large-bodied prey species, bush pigs (45–115 kg) and yellow-backed duikers (45–80 kg), make up a higher percentage of the animals hunted around Kabo. This shift towards smaller-bodied species around Pokola is highly indicative of hunting patterns in areas where larger-bodied wildlife species are depleted. Foraging theory also says that the prey set will increase as high-value animals become scarce (Stephens and Krebs, 1986), and the number of different species taken during the controlled hunt is greater around Pokola than Kabo.

Two factors probably contribute to the non-sustainable harvest of wildlife. First, continued population growth coupled with insufficient protein alternatives to bushmeat in logging villages means that demand for animal protein is higher than the supply of imported meat and fish. Thus, pressure on wildlife is high. Second, local communities still perceive wildlife resources to be unlimited and fairly easy to hunt (Lewis, 2002; Bowen-Jones et al., 2003).

Does protein availability meet the dietary needs of local people?

As previously mentioned, population growth associated with logging contributes greatly to pressure on wildlife resources. As of 2004, CIB employed nearly 1,600 permanent workers and 400 temporary workers, making it the largest logging operation in Congo. It pays approximately US$6 million per year in salaries, or about US$3,000 per employee (CIB, 2006), well above the average income per capita of US$650 (www.finfacts.com). Therefore, the average CIB employee has over 4.5 times as much income as the typical Congolese. There is a positive relationship between household income and protein consumption, with wealthier households more likely to consume greater quantities of animal protein (bushmeat, fish and imported meat) than less wealthy households (Auzel and Wilkie, 2000; Wilkie et al., 2005). In fact, bushmeat makes up a larger proportion of household diets for CIB employees than for non-workers (Moukassa et al., 2007).

To meet the protein needs of the population, PROGEPP partners have worked to increase the availability of alternative sources of protein in logging villages. Perhaps the most effective activity was the construction of five cold rooms and two butcher shops, resulting in the ability to store imported meat (pork, chicken and seafood) and to butcher imported cattle. Since 2002, an annual average of 79,200 kg (SD = 15,565 kg) of frozen meat and 29,772 kg

Box 9.4   Does the quantity of protein meet the requirements of the residents of the concessions?

By estimating the protein requirements of an average person, we can assess whether the current quantities of imported meat roughly satisfy the needs of concession residents. An average person requires approximately 1 g of protein per kilogram of body weight per day. Thus, assuming an average body weight of 40 kg (an average weight for children and adults in the tropics), the average human requires 40 g of protein per day and 14.6 kg of protein per year. As boneless meat contains about 20% protein, the annual requirement of meat is 73 kg per person. With an average of about six people per employee household, approximately 12,000 people depend directly on the logging company (CIB employees and their families), and together require a total of 876,200 kg of protein per year. Thus, the alternative protein imported into the concessions (~100,000 kg per year) provides at best 11.4% of the protein needs (876,000 kg) of CIB employees and their families and 4.8% of the needs (2,063,272 kg) of the entire population of the concessions.

What percentage of the overall protein requirement needs to be satisfied by imported meat versus freshwater fish and wild meat? For Kabo and Pokola, from which imported meat and market data are available, the overall requirement for protein sources of all the residents is 1,163,474 kg per year. Assuming that current rates of fishing are sustainable and that local people currently satisfy their daily requirement of protein, freshwater fish can continue to provide an average of 41% of household protein (477,02 kg/year). Note that for unexplained reasons freshwater fish only accounted for 26.5% and 12.1% of household diets in 2003 and 2004 in Pokola, and thus seasonal and annual fluctuations in fish abundance will largely influence consumption of other sources of protein (Moukassa et al., 2007). Robinson and Bennett (2000) derived a theoretical estimate of tropical forest productivity of 152 kg/km^2 and noted that off-take from several forests suggests that overall productivity is less than 200 kg/km^2. Thus, using the liberal estimate of productivity, the hunting zones around Kabo and Pokola are likely to sustainably provide 280,400 kg of wild meat per year. Therefore, based on the above estimations, the level of imported meat would need to be increased threefold to 406,050 kg to satisfy the protein needs of residents of Kabo and Pokola.

(SD = 10,440 kg) of beef has been imported into the four forestry sites (Pokola, Kabo, Ndoki 1 and Ndoki 2), yet these efforts remain insufficient to feed CIB employees and their families (Box 9.4).

Imported meat or intensification of aquaculture and animal husbandry will have to dramatically increase to ensure that residents of the concessions have access to sufficient protein while simultaneously preventing overharvest of wildlife populations in the concessions. Despite the gap between demand and supply of imported meat, the percentage of domestic meat in household diets (5–7.2%) increased significantly between 1999 and 2005 (Moukassa *et al.*, 2007), perhaps demonstrating that under the right conditions imported meat could serve as a substitute for bushmeat.

Alternatives to hunting as a source of income

A back-of-the-envelope calculation can provide an idea of the importance of bushmeat as a source of income for local people in concessions. Using only the quantity of wild meat arriving in the markets of four logging villages (Pokola, Kabo, Ndoki 1 and Ndoki 2), we multiplied the annual mass of bushmeat in each market by US$0.80 (400 CFAfr); this is a conservative estimate of the price of a kilogram of bushmeat received from a trader in a market and is based on the price of meat of the blue duiker, *Cephalophus monticola*. On average, over the last 7 years, the annual return of bushmeat in markets is estimated at US$103,692 (51,846,213 CFAfr) for the four sites (Table 9.1). Sale of bushmeat represents approximately 1.6% of the cash economy (total of CIB wages, PROGEPP wages and bushmeat income), spread among the hunters, gun owners and traders who play a role in the bushmeat commodity chain. Although this estimate of revenue is conservative because it does not take into account bushmeat sold outside markets, even if 50% of bushmeat is

Table 9.1   **Estimates of total biomass (kg) of bushmeat appearing in the markets of each of the forestry sites. The total cost in CFAfr is based on a conservative estimate of 400 CFAfr ($0.80) per kilogram of bushmeat.**

Year	Kabo bushmeat (kg)	Ndoki 1 bushmeat (kg)	Ndoki 2 bushmeat (kg)	Pokola bushmeat (kg)	Total bushmeat (kg)	Total* (US$)
1999	23,496	43,218	35,244	112,528	214,486	161,876
2000	21,716	37,519	27,056	57,494	143,785	108,516
2001	22,072	35,358	16,376	34,653	108,459	81,856
2002	18,156	25,191	18,156	56,946	118,448	89,394
2003	21,360	21,784	13,172	73,560	129,876	98,020
2004	16,020	11,093	11,036	65,949	104,098	78,564
2005	16,376	9,317	6,764	55,700	88,156	66,533

*Based on a conversion rate of US$1 = 530 CFAfr.

sold outside markets the overall sale would only account for 3.2% of the cash economy.

Although bushmeat sales are unlikely to contribute greatly to economic development, they could potentially provide a good source of income to a few indigenous hunters if the number of hunters was kept low and wild animal populations were sustainably harvested. In addition to being a food source, bushmeat has also been shown to be an important source of cash income for rural poor people in the Democratic Republic of Congo (de Merode *et al.*, 2004). This may also be the case for the people of northern Congo, particularly for the Mbenzele, who are less likely to have permanent jobs with the logging company than Bantus (Lewis, 2002).

One of the goals of the PROGEPP alternative activities programme is to develop activities (gardening, animal husbandry, fish production) that can earn cash as a substitute for commercial hunting. To date, none of these activities has earned enough money to substantially contribute to the monthly income of a typical household. Animal husbandry in Central Africa is complicated by the perception of domestic animals as savings and insurance rather than as sources of protein, and by parasites and disease (tsetse flies and trypanosomiasis) that severely limit large-scale livestock production. Early in the project considerable effort was focused on rearing of non-traditional livestock, such as snails, rabbits and guinea pigs. These experiments failed to be productive or were not embraced by local people. Efforts to farm fish were more successful, but they depended upon a large investment by the logging company in terms of providing a bulldozer and driver to dig the ponds. Of the alternative activities that PROGEPP has promoted, selling fishing supplies at low costs (prices of supplies in Brazzaville or Douala) may have an important economic return for local people. While helping local fishermen to intensify fish harvest may be an alternative to hunting, particularly given the importance of fish in local household diets, the effects of fishing on fish stocks need to be studied to ensure that one unsustainable activity is not being traded for another.

The road towards sustainability

The road towards sustainability of livelihoods and wildlife populations is an incremental one. At the start of the project in 1999, hunting camps lined the logging roads and the primary goal was to put an end to the poaching of protected species (Box 9.2). The socioeconomic context was also different, with 70% fewer people in the logging villages. Now, with the poaching of gorillas and elephants at a minimum, the goals of PROGEPP have evolved to focus on the management of game species as a food source for the residents of the concessions.

According to our analyses of market and duiker off-take data, the current levels of hunting around the logging villages of Kabo and Pokola are probably not sustainable. These results are based on extrapolations of data averaged over several years and estimates of protein requirements and forest productivity borrowed from the literature. Though not precise, these estimates paint a reasonable picture of the pressures on wildlife around the two largest villages of Kabo and Pokola. As an aside, it is important to note that the results depended on data specific to the concessions (harvest rates, market data and household consumption data). In the absence of data, the road to sustainability is a blind one, and additional information is necessary to refine management strategies required to prevent unsustainable harvest of wild animals.

Given that hunting is probably unsustainable around the two largest villages in the logging concessions, the question is how to alter the current management system to curb levels of hunting and guarantee adequate sources of protein for the concession residents in the long term. Project successes to date, coupled with continued existence of harvestable species in the concessions, suggest that this goal is achievable if project partners (CIB, WCS, MEFE and local communities) work together to achieve this end. We suggest the following as possible strategies to achieving sustainability of wild animals in the concessions:

1   *Import greater quantities of meat and subsidize prices.* While PROGEPP and CIB have made progress in ensuring the availability of frozen meat and livestock in the concessions, the amount of imported meat transported into the logging villages does not meet the demand of CIB workers and their families, let alone other inhabitants of the villages. In addition to inadequate quantity of imported meat, prices of imported meat are high relative to freshwater fish and bushmeat and must be reduced to be competitive (Moukassa *et al.*, 2007). Increasing the quantity of imported meat will probably depend on an initial intervention by the logging company, which has the logistical structure and means to start up large-scale importation of meat for its employees. At early stages, the company should subsidize meat importation, helping private operators transport the meat and keeping prices artificially low at a level competitive with bushmeat and fish. At similar prices imported meat may not serve as an effective replacement for wild meat unless the availability of wild meat is restricted (Wilkie *et al.*, 2005). By managing for wildlife (law enforcement and restricted access to wildlife) the quantity of wild meat in markets will be limited to sustainable levels, and imported meat might therefore be a reasonable substitute.

2   *Make alternative sources of revenue and protein available.* Although it is likely to take time, activities that supply alternative sources of revenue and protein must be made productive and profitable. Rather than experimenting with new types of animal husbandry or agriculture, using formal training

to improve techniques for activities (e.g. raising goats and sheep) that already exist in the concessions may prove more successful. In addition, working with locally organized village groups may be more successful than working with individuals, who may become uninterested or unable to give enough attention to activities due to other responsibilities.

3 *Empower local people.* In national law, people indigenous to an area of Congo and immigrants from other parts of Congo have equal rights to the use of natural resources. Adoption of a zonation system based on land use practices of indigenous people was a first positive step towards reinforcing local authority over their traditional hunting, fishing and gathering zones. However, progress must still be made in preventing management decisions in logging concessions from marginalizing indigenous populations. CIB employees still enjoy privileges not extended to non-workers because of their relatively greater wealth, their ability to organize themselves through worker unions and the simple fact that the logging company is interested in treating its workers well so that they will be productive employees. This puts non-workers at a disadvantage because they lack organization and representation. This is particularly true for the Mbenzele, whose lack of formal education, attachment to an 'immediate return' economy and forest lifestyle results in a lack of representation within the logging company and government (Lewis, 2002). Therefore, local people need to be further empowered through organization, capacity-building and sharing of information.

4 *Study the economy of hunting and sustainability.* The idea of a managed system is to produce the maximum output while not risking long-term sustainability. It is therefore important to closely monitor game populations and to improve our understanding of the demographics of game species in order to determine the maximum harvest possible. In addition, an explicit study on the hunting commodity chain should be conducted to provide more detailed information on the level of hunting, off-take per effort, final destination of bushmeat and the net profit incurred from hunting for each of the actors along the chain.

The results of this study are largely context specific. No other logging concessions in Central Africa have benefited from the level of investment and effort towards wildlife management as those concessions allocated to CIB. That this system is an exception to the norm is also evidenced by the recent adoption of the Kabo management plan and the company's commitment to Forest Stewardship Council (FSC) certification, demonstrating CIB's commitment to responsible and sustainable management of forests. However, this approach to conservation can be generalized and replicated. We strongly urge logging companies and governments to estimate sustainable levels of wildlife harvest, and then find ways to make up the difference – at minimum guaranteeing employees and their families a constant and adequate supply of animal protein at prices equivalent to bushmeat and freshwater fish. As most logging operations depend on the immigration of workers into isolated forests, logging companies have a responsibility to feed their workers. Relying on bushmeat as the only protein source means that the logging companies are essentially profiting from two resources – tropical timber and tropical wildlife. A commitment to the livelihoods of local people and sustainable management of wildlife will require an investment by the logging company.

Wildlife management and conservation have been criticized for prioritizing animals over local people and their cultures. In so far as the livelihoods of local people are tied to wild meat, however, the goals of wildlife management coincide with development goals when the unsustainable harvest of wildlife could put food supply at risk (Bennett *et al.*, 2002; Davies, 2002; Fa *et al.*, 2003). Moreover, the depletion of wildlife and habitat degradation have been blamed for making semi-nomadic forest people sedentary and for their loss of cultural knowledge (Lewis, 2002). Therefore, wildlife management has the potential to contribute to food supply and conservation of the culture of forest peoples, although its limits should also be recognized. Bushmeat makes up a large proportion of the protein in household diets, but it is unlikely to contribute to the alleviation of rural poverty or to significantly increase household incomes. Even if hunting can be managed for sustainability, it is unlikely to maintain the livelihoods of the entire population of the concessions. Although bushmeat may be abundant enough to act as the principal protein source of indigenous people, immigration for logging pushes the demand to exceed the supply. Bushmeat has contributed, and can continue to contribute, to the diets of local people. The challenge is to determine who should benefit from this limited resource and how to guarantee the livelihoods of local people without depleting wild animal populations.

Acknowledgements

We thank the government of Congo and the MEFE for its support of the work described here. The governments of Switzerland, Japan, the USA and France, ITTO, USAID, CARPE, USFWS, USFS, WCS, CIB, the Liz Claiborne Art Ortenberg Foundation and Columbus Zoo have contributed to the funding of the project. PROGEPP and the activities described in this chapter have benefited from the ideas and work of a large number of people. We would particularly like to thank the following people for their contributions to the project and wildlife management in the concessions: P. Elkan, S. Elkan, P. Kama, A. Moukassa, R. Malonga, M. Zoniaba, B. Kimbembe, C. Makoumbou, C. Prevost, P. Mbom, D. Paget, O. Desmet, P. Auzel and J.-M. Mevellec. Comments by D. Wilkie, D. Poulsen, G. Davies, D. Brown and E. J. Milner-Gulland greatly improved this manuscript. Special thanks are due to the local communities that have supported the work and to PROGEPP employees and the ecoguards who are the foundation of the project.

10

Institutional Challenges to Sustainable Bushmeat Management in Central Africa

Andrew Hurst

Introduction

The problem of unsustainable bushmeat exploitation in Africa has received widespread attention recently, nowhere more so than in the 'bushmeat heartlands' of Central Africa. Several governments and organizations in Cameroon, Gabon and the Republic of Congo have attempted to solve the problem by addressing the needs and perspectives of those people most dependent on bushmeat: the rural poor. Three projects in the region stand out as constructive efforts: the Community Hunting Zones initiative of the Ministry of Environment and Forests of the government of Cameroon; the Minkébé Project led by WWF Gabon; and the Project for Ecosystem Management in the Periphery of Nouabalé–Ndoki National Park (PROGEPP) led by the Wildlife Conservation Society (WCS) in the Republic of Congo.

Each of these projects is unique in its attempt to account for local socioeconomic dynamics in conservation efforts – and by extension the articulation of local and national political economies. A closer examination of the context of bushmeat exploitation as confronted by these projects can provide insight into the nature of the bushmeat challenge and, in turn, may help improve attempts to meet it. Doing so may also provide some general insight into common factors that policy-makers and donors, as key players at the national level, must account for in their efforts to support conservation initiatives that impinge upon the livelihoods of local people.

Institutions and the governance of bushmeat exploitation

The main finding of this chapter is that, for conservationists to meet the bushmeat challenge, a better understanding is required of the social, political and economic dynamics within which bushmeat exploitation takes place. International conservation efforts are too often predicated on implementation taking place in a geographic vacuum. In the case of social relationships, those working on conservation projects, whether they are expatriate or local staff, cannot avoid influencing, altering or in some cases supplanting the human relationships that sustain the 'problem' that projects are designed to 'solve'. Most directly, there are relationships established with the communities at the 'front line' of the issue. Yet relationships are also established with local markets for goods and labour. Equally, there are almost always relationships established with state actors at the local and regional levels.

Bushmeat exploitation is embedded within relationships like these. Reflecting on the context forces us to consider certain questions: Why is it that only certain individuals hunt? How does the position of these individuals in the local community influence hunting patterns? Are there social relationships that extend beyond the local area which sustain hunting as an activity? If so, what are they and do they extend beyond bushmeat into other areas of the local, regional or national economy? Answering these and other questions and thinking through the nature of such relationships helps identify the interests of the actors that form these relationships and the norms according to which they operate.

In fact, most conservation projects, whether wittingly or not, are designed to alter these relationships. Some activities simply supplant existing relationships, as when a conservation organization provides a job as an animal tracker to a former hunter once dependent on a patron for a gun and payment for bushmeat. Other interventions are designed to eliminate certain relationships, such as when the link between a rural hunter and an urban consumer is severed by eliminating the supply of bushmeat to urban areas. Conservation projects that attempt to influence, alter, replace or eliminate relationships in this way can often have both intended and unintended consequences. Local markets for goods or labour can be distorted. Local officials can be relieved of the responsibility of performing functions their mandate requires them to fulfil. The consequences of conservation projects thus have a bearing on the effectiveness of the interventions themselves.

Of equal importance are the consequences for the wider context of governance and state functioning in countries with weak state institutions and typically low levels of responsiveness to the needs of the poor. As a major consumptive resource that provides benefits to the poor in many developing countries, bushmeat has importance beyond conservation. Understanding the context of bushmeat exploitation is important because it allows us to consider conservation interventions as part and parcel of the social, political and economic context. This in turn allows us to consider more generally the effectiveness of the arrangements that place non-government actors of external origin at the centre of bushmeat management in the humid tropics.

Profiles of bushmeat management interventions

On behalf of the Overseas Development Institute (ODI), the author undertook a short research visit to the Republic of Congo, Gabon and Cameroon in late

July and early August 2004 to examine the institutional challenges faced by these three projects.[1] The three different projects are outlined briefly below.

Community hunting zones in Cameroon, government of Cameroon and WWF Cameroon

Legal provision was first made for community hunting zones in Cameroon in the revised National Forest Policy of 1994. The first community hunting zones established drew on the community forest model, were limited to 5,000 ha in size and were permissible only in the non-permanent forest estate. This made sense in the savannah, where large populations of wildlife could be found in quite small areas. But in the humid forest zone, hunters habitually covered territory 10 times as large, owing to the dispersed nature of forest animals. As a result, WWF Cameroon began to lobby for a community hunting model better suited to the humid forest zone. In response, the government created a new type of community hunting zone in early 2001, called Zones d'Intérêts Cynégétiques à Gestion Communautaire (ZIC à GC). These new community hunting zones were envisaged to be upwards of 100,000 ha each. With the help of WWF, the government mapped out 13 such zones in the south-east where inventories carried out indicated that wildlife could be harvested sustainably if a suitable management plan was in place. Zones were delineated on the basis of administrative boundaries. The law stipulated that community hunting zones could be established in the permanent forest estate and could overlap with forest concessions.

The government has handled applications for community hunting zones at the local level through the departmental delegate of Le Département de la Faune et le Chasse[2] (Wildlife Department). Central representatives of the department have been actively promoting the zones in cooperation with WWF. In order to take over the management of a community hunting zones, the government requires communities to establish a local bushmeat management committee. Communities are eligible only if they are within one of the 13 hunting zones mapped out by the government. To date, the ministry has agreed to the community management of five ZIC à GC; three of these are currently operational. For its part, WWF has done awareness-raising in the eligible villages in order to explain the rules and procedures It has also provided support to communities to develop management plans and to carry out wildlife surveys in their areas. At the time of the field visit, one inventory had been completed.

Communities are encouraged to sell some of the products of the hunt to generate revenue. Communities are also allowed to lease out their hunting zones to commercial trophy-hunting operators. A portion of the income received from these leases must be paid to the government. The percentage is not fixed in advance but negotiated between the government and the local communities. In addition, taxes paid by the safari operators are divided between local communities and various levels of government in the same manner as the forest exploitation taxes: local communities receive 10%, the local commune receives 40% and the national treasury receives the remaining 50%. The use of community funds is overseen by a village development committee, the same body responsible for overseeing income from the forest exploitation taxes where they are earned.

The law as it stands supports the distribution of costs for wildlife management. Most new community hunting zones overlap with forestry concessions, and every forestry concessionaire is supposed to carry out a wildlife inventory that could in theory serve as a baseline against which sustainable off-take levels could be established. Although there is an asymmetry of power between communities and trophy-hunting outfits, the WWF has acted as an intermediary, helping communities negotiate and manage the relationship and to build capacity for the management of wildlife under their control.

Policy contradictions currently exist in the permit systems for hunting and meat transportation, and these provide disincentives for the legal sale of bushmeat from community hunting zones. Technically, Cameroonian law requires a permit to transport bushmeat. However, most hunting zones do not have these permits. Moreover, the permit does not require those in possession of them (typically market sellers) to identify where they obtained their bushmeat. This creates an obstacle to the promotion of 'sustainably harvested' bushmeat. The Cameroonian government is working to resolve this problem, in cooperation with WWF, by creating a system that will allow community hunting groups to sell their meat legally in town. The government is considering changes to the law that would oblige traders to purchase bushmeat from 'sustainable' sources, which in effect would mean the community hunting zones. According to WWF, restaurants in Yaoundé, for instance, are already in direct contact with communities managing community hunting zones. In addition to legalizing potentially lucrative income streams for local communities, such changes would make circumvention of the rules more difficult by attempting to control secondary bushmeat markets – markets that should be easier to regulate and control than the highly decentralized primary bushmeat markets.

The Minkébé Project, WWF Gabon

WWF initiated its project in the forest concessions around Minkébé National Park in 1997. The project covers 6 million ha including the park. To date, WWF has facilitated the adoption of a protocol, in March 2001, in one of the big concessions between the Ministry of Water and Forests, the Ministry of the Interior through the governor of Woleu-Ntem province, two timber companies and four villages of Elelem Canton. The protocol forbids hunting by forestry workers within the concession. Companies agree also to not allow bushmeat to be transported on company vehicles. In addition, WWF supports surveillance of the park and forest concessions with mobile patrols of ecoguards. Ecoguards are officially staff of the Ministère des Eaux et Forêts, but they have contracts with, and are paid by, WWF.

According to WWF studies in the area, bushmeat hunting is generally limited to within a 20-km-wide corridor near the Ndjole–Oyem road, as this is about the maximum distance that a hunter can walk in a day. Organized hunters often use old logging roads to venture further into the forest, setting up camps and spending several days hunting for bushmeat, ivory and skins. A large part of the hunt is for unprotected species, estimated to be 80% of the off-take, although WWF does not have any studies indicating exact numbers. However, protected species can be found in the local markets.

Most people in the area are Fang, although Baka pygmy are a significant minority. Both groups are involved in hunting. Some Baka live permanently in the forest and generally barter for goods in return for hunting. Other Baka are more settled and live in villages, participating in the cash economy. There is also a large immigrant population. In Oyem, for instance, there are over 300 Malians involved in trade, running shops and driving taxis. From experience, WWF believes that elephant poaching for ivory is generally driven by people from Cameroon and Equatorial Guinea, who hire Baka pygmies to do the hunting.

There are a total of 15 companies working concessions around the National Park, and they include French, Lebanese, Libyan, Malaysian and Chinese interests. The proximity to a good road and nearby towns means that companies have to date not provided alternative protein sources for their workers. At one of the concessions, the manager explained that there were two shops available for workers to purchase food, both under contract to the company.

The complicated corporate environment is accompanied by poor state capacity on the ground. The Ministry of Forests has few staff in the field. Staff responsible for managing hunting at the ministry headquarters in Libreville complained of insufficient resources. This may be true (there are fewer than 40 people in their division for the whole country), but it is probably a result of the low importance given by the state to hunting and therefore to low-value resources of interest only to the poor. It is also indicative of the concentration of wealth and social opportunities in Libreville, making it difficult to keep bureaucrats in field positions. There are government representatives with regional responsibilities based in Oyem. In addition, district-level officials are also based there; their time is split between administration (licences for hunting and logging) and visits to forest concessions to ensure compliance with various government measures. There is only one vehicle in Oyem, one vehicle in Makokou and no vehicles in Mitzic. Officials receive per diems and transportation from the logging companies they are meant to supervise.

The Nouabalé–Ndoki Buffer Zone Project, WCS Congo

WCS has run PROGEPP in the forest concessions of northern Congo for the past 4 years. The project has two principal aims: to increase the land base for conservation and to work with logging companies operational around the national park to improve their environmental management. There are four large concessions around Nouabalé–Ndoki National Park:[3] Kabo (300,000 ha), Pokola (560,000 ha), Loundougou (386,000 ha) and Mokabi (375,000 ha).[4] All four concessions are held by CIB. Prior to the arrival of the logging company in the 1970s, there were no roads in the area beyond Ouesso and the majority of residents in the area's villages or settlements were Baka pygmies. With the arrival of the logging company, the non-pygmy population increased. Pokola, originally a fishing village of 400 inhabitants, has grown to a small town of 12,000 people. Other logging areas with settlements exist at Kabo, Ndoki 1 and 2 and Loundougou. Approximately 80% of CIB staff originate from outside the area, representing 50 different ethnicities. The remaining 20% are pygmy and non-pygmy peoples already resident in the area. A census in Pokola in 2002 indicated that about 6% of the population was pygmy (Baka, Benjele and Ngombe).

Currently, commercial hunting in Congo is illegal although two kinds of non-commercial hunting are allowed. 'Traditional hunting', defined as the use of traditional methods for subsistence use, is allowed at any time of the year. 'Hunting with a rifle' is only allowed during the 6-month hunting season that begins on 1 May and ends on 31 October.[5] A permit must be purchased annually from the Ministry of Commerce, Forestry and the Environment in order to hunt during the season. Permits technically apply only for a particular gun and a particular person, but this is not strictly enforced, as most people who own guns do not hunt themselves, sending others out into the bush to hunt for them. Hunting with metal cable snares is illegal. Pygmies make up the majority of hunters in the area.[6] The majority of hunting by pygmies is done in the context of patron–client relationships with non-pygmy families living in Kabo and other areas. WCS studies indicate that the majority of bushmeat consumed in households in the area is purchased at the market.

WCS staff work in conjunction with officials from government and CIB to enforce existing laws and company regulations, to educate and raise awareness amongst school children, loggers and the general population of the wildlife law and the benefits of conservation, and to promote alternative sources of protein and income for local people. Ecoguards are technically hired by the ministry and receive a base salary plus incentives for seizing illegally hunted bushmeat, guns and snares.[7] CIB contributes approximately 15% of total project costs, with the majority of the contribution going towards law enforcement (ecoguard salaries etc.) and the remainder supporting conservation-related research in the concessions.

For its part, CIB has attempted to offset bushmeat demand amongst its workers in Pokola by supplying beef at subsidized prices. Cattle are purchased live from Cameroon or the Central African Republic, and brought to Pokola, where there are butchering facilities and four cold storage units. Two other cold storage units exist at Kabo while Ndoki 1 has one as well. Beef is butchered into small portions to reduce the likelihood of it being smuggled to Ouesso, where it could be resold for a premium.[8]

The project is moving into a new phase by attempting to create a land management system based on a 'patchwork mosaic of 'sources and sinks', with hunting and no-hunting zones for both villagers and for loggers. Six hunting zones are to be established next to Pokola, with access rotating monthly. Loggers will be allowed to hunt in a zone on their day off. The plan was designed to be included within the terms of the management plan that CIB is required by law to prepare. If the government accepts this, then these zones will become law.[9] The proposed zones were established through consultations with local representatives and in cooperation with government officials. Once areas were agreed, an agreement was signed by the village head.

The institutional challenges of bushmeat interventions in Central Africa: reflecting on policy and programme frameworks

The three initiatives outlined above all touch on the institutional dynamics of bushmeat exploitation in both intended and unintended ways. Different as they may be, some common dynamics of the institutional context of bushmeat exploitation can be found, and some common concerns are raised about the effectiveness of bushmeat conservation projects and the policy frameworks that inform them. In light of more general trends in post-colonial African governance, the experiences of these projects suggest that new strategies may be required to achieve sustainable bushmeat exploitation.

State institutions

State institutions are powerful determinants of the patterns of resource exploitation and, because of policy contradictions, they often operate at cross-purposes with each other. State institutions also frequently enjoy vastly different power bases within government. The relative power of competing ministries often depends on the relative priority of particular agendas within government or amongst supporting donor agencies and countries.

The generalized trend to decentralize governmental structures in Africa provides another challenge for sustainable bushmeat management. Decentralization in practice has meant greater power for local authorities, but this has rarely included an adequate increase in their resources. Field staff are generally poorly equipped and meagrely paid, and even the most dedicated and committed suffer from low morale. The lack of adequate of resources can leave field state officials reliant on others for transportation and create incentives for bribery and other forms of corruption. State-led efforts to control bushmeat exploitation in rural or remote areas are therefore impeded from the start, and are at risk of being undermined by the very people who are supposed to be the agents of good governance. Rent-seeking behaviour or co-optation into bushmeat-related social networks often results. The prevalence of these kinds of relationships also makes it difficult to ensure that the original conservation objectives survive the distortions of local power politics. WWF in Cameroon approached this challenge by working with the departmental delegate, a person in a position of authority 'high enough' to avoid serious problems with corruption, but who can still be overseen by provincial delegates to ensure this. In Congo, WCS tried to ensure a significant presence on the ground of staff able to support local government capacity and/or limit opportunities for corruption and subversion of project aims. Even so, historical perceptions are difficult to overcome, and it will take some time to engender respect for state regulations amongst local people, many of whom consider petty corruption endemic to local authorities.

To a certain extent, the arrival of conservation NGOs has had positive effects on bushmeat exploitation, by altering power relations. Where exploitation pressures have been generated by activities of other departments (such as increased access to the forest through forestry activities), NGOs can supplement the power of those officials mandated to control the bushmeat trade. In the unusual case of northern Congo, WCS and CIB have essentially supplanted the state by assuming its command and control functions. In Gabon, WWF supports the government in policing bushmeat harvesting. Yet such situations often lead to a blurring of roles and responsibilities. The WWF ecoguards in Gabon present a rather confusing picture to forestry workers and local villagers. Employees wear uniforms that have an Eaux et Forêts patch on one sleeve and a WWF patch on the other. Patrols are conducted in a vehicle with a WWF logo on the side but no government insignia. Nor is WWF a party to the protocol which spells out responsibilities between the concessionaire, the government and local villages, despite working with both the concessionaire and the government to enforce the protocol. While such arrangements may work in the short term, they cannot prevail in the longer term. If states are ever to accept ownership of the problem, and if a solution is ever to be found that ensures a modicum of accountability to the local people who depend on these resources, then arrangements for bushmeat management must include the meaningful involvement of state officials.

Private enterprise and commercial institutions

Commercial institutions have a very obvious impact on unsustainable bushmeat exploitation. Addressing their role in sustainable bushmeat exploitation is arguably the easiest institutional challenge because they operate according to a clear and singular principle: the financial bottom line. As long as forestry companies can make a profit, they are likely to be receptive to the idea of controlling bushmeat exploitation in their concessions.

While the approach taken by WCS Congo in engaging with CIB is laudable, the situation is unlikely to be replicated elsewhere. CIB depends on the European market for over 90% of sales. All the roads in its concessions are private, and therefore eminently controllable. By contrast, in Gabon, there are 15 different concessionaires around the park, and soon to be more. The situation is even harder in south-east Cameroon, where there are few concessionaires but many more operators. Some companies in south-east Cameroon are cooperating with anti-poaching efforts and providing alternative protein sources for their workers, but they seem to be having little impact on their local competitors. Figuring out how to engage with the large number of smaller operators – whose margins lessen their interest in absorbing social costs in their operations and whose political connections make it easier for them to sidestep regulation – remains a challenge.

With the larger concessionaires, there is some merit in the idea of factoring the management of bushmeat exploitation in forest concessions into certification schemes, particularly where their markets so demand. But in global terms, the forestry industry has a history of mining forests and then moving on. Its corporate leaders therefore cannot necessarily be relied upon to lead. And given that China, one of the world's largest consumers of wood, is not presently a green market, certification in consumer countries is probably not the only answer.

The involvement of trade unions may also help to increase the responsiveness of logging companies in Central Africa to unsustainable bushmeat exploitation. For instance, in Gabon, most workers in forestry concessions are members of one national union which, along with its members, could be a useful counterbalance to logging companies in addressing the needs of forestry workers. The unions may have a role to play on issues such as alternative protein sources, codes of conduct or inducements to limit hunting by unionized workers. It could also be an entrée into national-level discussions over future agricultural and food provisioning strategies. WCS Congo has done exactly this. PROGEPP is engaged with labour unions, and their representatives have put pressure on CIB to provide alternative sources of protein. At the same time, the union is working with its membership to encourage cooperation and compliance with project activities.

The politics of markets and social institutions

The most difficult kind of institutions faced by conservation organizations working to manage bushmeat exploitation are those informal institutions that operate at the cross-section of the market, the formal political system and the myriad social networks. These informal institutions tend to transcend local boundaries by linking elites in regional centres or cities with communities or villages in bushmeat source areas. They are often the source of the status, influence and authority of elites, be they local, regional or national.

All three NGOs have recognized the role that these networks play in bushmeat exploitation and have attempted to circumvent them through management strategies that distinguish between subsistence and commercial hunting. Subsistence hunting is defined in geographic terms (meat can be traded only in the local area), effectively making the boundary between legitimate and illegitimate resource claims territorially bounded. Yet the distinction is not necessarily a clean one. When does someone cease to be local? Are students away at school no longer local? What about those who have left the area temporarily to work? Are they only local when they return to the area?

Moreover, are there ever any instances when external interests might have a legitimate interest in bushmeat?[10] In Congo, hunting zones are granted to forestry workers, who are mostly from outside the local area. But one informant noted that this may sit badly with local people, who already see little benefit from the presence of the logging company in their area. Logging company employees already receive a good wage and might not need to benefit from free access to local bushmeat through hunting zones for their exclusive use. As an alternative, such hunting areas could have been allocated to local villagers, who could then sell bushmeat to logging company employees and into the logging town.

In fact, such distinctions between subsistence and commercial hunting do not accurately capture the character of the social institutions that are involved in bushmeat exploitation. Institutional norms mean that individuals define the legitimacy of resource claims according to a different notion of proximity. Legitimacy depends more on kinship or political ties to an area than it does on one's physical presence there. At the same time, local hunters do not define entitlement according to where a resource is consumed. This is why it is so difficult to patrol the 'borders' of the areas within which 'subsistence' hunting is allowed. To designate a resource such as bushmeat as illegitimate because it has been 'commercialized' (i.e. it has been transported out of the local area) confounds many existing norms of entitlement.[11]

One alternative frequently mentioned is to legitimize access to a resource solely on the basis of customary rights. But this too is difficult and can contribute to local conflict, especially when customary or traditional claims to resources are contested and local definitions of who is and is not 'indigenous' change depending on the context in which the term is used. These points may be less important now in areas that are quite remote and have limited out-migration. But as road-building increases and transport becomes easier, they will surely increase in importance. External efforts to define legitimacy via distinctions between 'subsistence' and 'commercial', or between 'traditional' and 'non-traditional', cannot easily overcome local understandings of legitimacy generated by the local social institutions in which bushmeat hunting and trade are currently based.

What other suggestions can be offered? Progress may depend on a more nuanced approach to social networks and resource entitlements. In Gabon, for instance, where hunting with a permit is legal, the government is considering regulating the bushmeat trade through the establishment of village associations. These associations would provide the only legal source of meat for outside traders and hunting could be done only by hunters registered with the village association who possessed the appropriate permits. This would build on existing accountability mechanisms, and may incorporate measures that buttress the ability of local people to avoid elite capture or negotiate more equitable relationships with elites. Urban consumers continue to have access to the resource while local people are ensured a fair share of the benefits to which they are entitled.

Functioning markets are often good ways to allocate scarce resources, and the approach taken by the government and WWF in Cameroon may represent another good approach. But, to be ultimately successful, market frameworks must be sufficiently flexible to allow suppliers and 'consumers' to negotiate equilibrium prices within the limits of sustainable off-take. Poorly regulated markets for hunting zones can lead to large profits for middlemen and inequitable returns for local people. Should such a situation arise, there is a danger that local communities will simply revert to poaching outside their legal hunting zones to make up for 'lost' revenue.

Conservation NGOs as institutional players

Each of the initiatives described above involves an international conservation organization. While international NGOs have been key forces in addressing bushmeat exploitation in Central Africa, one should not ignore the fact that they too are part of the institutional context. Their presence affects patterns of bushmeat management and exploitation. This is most obvious when competition exists between organizations working in the same country or region. Such rivalries can hinder the cooperation and collaboration that are essential when the problem must be managed at scales too large for any one organization.

International NGOs are also part of the institutional context as players in local labour markets, sometimes competing with governments to hire high-quality staff. This is especially so where they are able to offer higher salaries than the public sector. In some instances, the reverse is true. In Gabon, WWF recounted how it is difficult to hire university graduates as they expect the high salaries and perquisites offered by government. Nevertheless, the presence of NGOs in labour markets is seen as positive by many. Staff are trained and given opportunities that are in scarce supply in weak economies. WWF Gabon often hires high-school graduates, who are provided with on-the-job training, thereby increasing the pool of skilled people.

In recognizing themselves as part of the institutional context in these countries and regions, conservation organizations must also better address questions of accountability. To the extent that NGOs are accountable, it is often to their northern constituencies, which are imbued with environmentalist views quite different from those of the forest dwellers and rural people whose lives are most affected by the projects and programmes. There is thus often a lack of formal accountability, either to the governments in whose countries they work or to the people in whose areas they operate. The power that NGOs derive from their northern constituencies (both governmental and non-governmental) means that African governments are often reluctant to challenge them. More fundamentally, there is little recourse for a villager if he or she feels aggrieved by the actions of an NGO in his or her area.

These issues come to the fore when considering NGOs' self-appointed role as a third-party verifier. In the case of PROGEPP in Congo, WCS staff see third-party verification as a key strength of the project, ensuring that CIB makes allowances for conservation within its concessions, as it is required to do by law. And, yet, local informants expressed concern that the process was insufficiently responsive to local needs. When government officials floated proposals to commercialize bushmeat, they were met with fierce opposition from conservation organizations, which were able to suppress the idea before it could even be discussed in any detail. This does little to foster ownership of policy by the government officials who are suppose to implement it and fuels criticism that conservation NGOs are only paying lip service to the principle of partnership. The point in this instance is not whether commercial hunting is or is not viable, but whether or not it was ever properly considered as a possible policy solution, evaluated accordingly, and then rejected or accepted on the basis of sound evidence. The fact that distrust of international NGOs exists makes it difficult for true third-party verification to occur. Third-party verifiers need to verify commonly agreed principles, not the principles of one particular party to the agreement.

General conclusions

Unsustainable bushmeat exploitation is much more complex than simply controlling hunters who hunt too much and consumers who eat too much. Both of these activities must be situated in the wider social context in which they take place. The kinds of social sanctions, political pressures and economic incentives and disincentives that drive exploitation patterns must be analysed and understood. Institutions, as 'sets of formal and informal rules and norms that shape interactions of humans with others and nature' (Gibson, 2001), are a central feature of what shapes the exploitation of bushmeat. They constrain some activities, including the kinds of local behaviour that might feed into conservation goals. They also aggravate others – in this case, the very levels of exploitation that are the international cause for concern.

Conservation organizations ignore the institutional context of bushmeat exploitation and management at their peril. As mentioned elsewhere in this volume, bushmeat is one of a class of goods informally conceded by range state governments to their rural citizens and, as such, tends to be left out of national-level political processes and policy dialogues. Unless this situation is acknowledged, bushmeat management efforts risk ignoring the very institutions that drive unsustainable exploitation. Interventions that accept and promote a division between goods consumed locally versus those consumed 'nationally' (to the extent that they appear in national revenue streams and are subject to national policies) risk a major error of perception. Ignoring these institutions will lead either to the development of policy that has no bearing on the realities of local political economies or, worse, will sustain the persistence of such institutions in parallel form, outside the formal governance of national laws and policies. Without linkages to national governance that match the national scope of the institutions of bushmeat exploitation, conservation efforts in the region will be unsustainable. Such a situation is undesirable for local people, the wildlife they depend on, and those dedicated conservationists who seek to secure the survival and flourishing of both.

Notes

1   In all three countries, he met with project staff, government officials and other stakeholders. In the case of the projects in Gabon and the Republic of Congo, he spent time visiting project sites and talking with individuals in the project's vicinity. The trip was by no means a formal evaluation or in-depth assessment, but it provided enough of an understanding of these interventions, the context they operated in and the wider political economies of the countries concerned to serve as a basis for assessing the institutional challenges faced by bushmeat management projects in Central Africa.

2   There is one delegate in each department.

3   Nouabalé–Ndoki, originally a concession, was gazetted as a national park in 1993 because of significant populations of animal species of conservation concern, including forest elephants, gorillas, chimpanzees and bongo antelopes.

4   Pokola and Kabo have been worked since the 1970s. Loungdougou and Mokabi came into production in 2002.

5   This period corresponds to the reproductive season in the south of the country.

6   In contrast, estimates are that about 15% of CIB employees engage in hunting themselves.

7   With incentives, salaries for ecoguards average 125,000 CFAfr per month (approximately 1,000 CFAfr to US$1.9). This compares with mill workers/loggers, who can receive up to 250,000 CFAfr per month with production incentives.

8   As of July 2004 in Pokola, 1 kg of beef with the bone cost 1,650 CFAfr. Smaller portions of 100 g on the bone were selling for 175 CFAfr; without the bone, 100-g portions were 200 CFAfr.

9   As it stands, communities already have the right to hunt for subsistence within UFAs although they do not have the right to cut wood for commercial purposes.

10 NGOs as expressions of external interest are discussed below.

11 Interestingly, opinion amongst some NGO local field staff appeared to contradict somewhat the distinction between subsistence and commercial use of bushmeat. In their minds, the distinction between subsistence and commercial hinged on the size of sales (a few duikers versus a whole truckload for instance) as opposed to whether bushmeat was consumed locally as opposed to being sold for consumption elsewhere.

PART 3

Extrasectoral Influences
and Models

Jo Elliott

This section of the book focuses on extrasectoral influences and models, linking strongly to the previous sections. An important issue brought up in the previous sections, and at the conference, was the status of bushmeat in human livelihoods. Bushmeat is increasingly being seen by conservationists as a livelihoods issue, but how firm is the foundation of this belief? Is it merely an attempt to access poverty-driven funding sources in an opportunistic way? Or does it reflect a real recognition that animal conservation and human welfare are strongly linked?

At one level, the conference indicated a need to maintain a separation of the two topics. The conservation lobby's primary concern with the bushmeat issue is the threat that the trade poses to (non-generalist) primates and other vulnerable species, particularly when conducted in protected areas and timber concessions in primary forests. Development practice, on the other hand, is focused on poverty reduction, so that investment of development agencies in the bushmeat issue is justified only if bushmeat is a source of protein and livelihoods for a significant number of poor people and the unsustainability of the harvest is a real threat to their welfare (this assumes that the poor are not able to adapt to decline in the availability of bushmeat over time). A key question to be asked is what is a 'significant' number of people in such contexts? Generally the data presented in the conference and book case studies were drawn from fairly small geographical areas.

The Part 1 case studies appear to indicate the following broad conclusions:

- Hunting in most areas is driven by lack of jobs and alternative livelihood options, as well as being a 'hungry-season' and emergency source of income.
- Hunting tends to be done by the poorest and those for whom it presents the least time and opportunity costs (though some hunters are specialists and are not primarily engaged in agriculture).
- Most poor communities are protein deficient and will switch readily between protein sources.

- In some areas, the keeping of domestic livestock is not a realistic option.
- Reducing poverty will generally reduce bushmeat consumption and hunting.

Part 2 called for rights-based approaches and appropriate institutional responses, with interesting presentations outlining examples of such responses. In Chapter 7, David Brown called for the debate to be taken beyond the wildlife subsector, and this is the focus of the following Part 3 chapters.

The CONACO project, presented by Dale Lewis in Chapter 11, is a particularly interesting illustration of the section theme. This project directly addresses the need to improve the livelihoods of poor communities living in wildlife-rich areas in Zambia's Luangwa Valley by providing access to improved agriculturally derived cash incomes. The project works with 12,000 poor farming households, organized into producer groups, to help them produce, package and market their non-cotton cash crops. The effect of increasing livelihood security appears to have had a direct and positive effect on local wildlife populations, as evidenced by the voluntary handing in of guns and snares and the independently verified increase in the success of local safari-hunting companies in terms of both number of trophy animals and reduced time needed to find trophies.

Experience with other NTFPs also commends itself for study in the search for extrasectoral learning. In Chapter 12, Elaine Marshall presents findings from research into the commercialization of plant NTFPs in Bolivia and Mexico. This research identifies the factors driving the livelihood impacts of the commercialization of a variety of NTFPs, from mushrooms to weaving products. The case study includes a review of the extent to which the poorest households participate in trade, concluding that the poorest are often disadvantaged or excluded from the trade – for example, because of the high capital costs of entry and/or lack of market information. Overexploitation is less of a threat to the products being reviewed than is land conversion. Legalization of the trade appears to reduce profitability, particularly the supernormal profits enjoyed by some entrepreneurs.

Chris Dickson and Neil Bird, in Chapter 13, present findings of a study examining the extent to which bushmeat and livelihood linkages are recognized or addressed in what is now the main planning instrument of international aid – the poverty reduction strategy paper (PRSP). They examine the treatment of forestry issues, particularly bushmeat, in a sample of PRSPs. They conclude that bushmeat is not high on the political agenda in most of the countries assessed, although forestry as a whole is better represented where it is a key economic sector. Bushmeat is not generally viewed as an economic activity, both because of the general illegality of the harvest and because of the lack of national-level statistics on its place in the economy. Where it is mentioned at all, hunting tends to be presented in PRSPs as a problem to be overcome rather than a valid component of the economy and a valuable protein source.

Though from a markedly different ecosystem to the other papers, Ross Thompson's case study of barren-ground caribou in northern Canada (Chapter 14) is highly suggestive of one way to improve the management of hunting for the bushmeat trade. The chapter documents the way in which, in the early 1980s, the approach changed from one of scientific expert-based control to a more participatory form in which aboriginal peoples held the majority on the management board. While researchers have become understandably sceptical of 'bottom-up' management as a universal remedy to the ills of development, the proof in this case is in the positive effects on the condition of the herds, which are now at high levels in historical terms. While Thompson is mindful of the challenges that lie ahead, he argues a strong case for an approach that is more responsive to the knowledge of the users of the resource, and to their interests.

During the conference, the discussion following the presentations was mostly centred on the Zambia work of Dale Lewis and the CONACO project, which stimulated considerable interest. The 'hopeful model' this presented, in terms of implications for coexistence of wildlife and agriculture, is very encouraging. However, earlier presentations at the conference had emphasized the lack of agricultural (crop or livestock) opportunities in extensive parts of bushmeat producing areas. These same areas (particularly remote forested areas) also offer little in the way of non-consumptive wildlife uses such as wildlife tourism. Hunting for bushmeat may well, therefore, be the sole economic activity with any real profitability in such areas, and this limits the options for extrasectoral development.

This section proved to be a useful opportunity to learn from and compare experience of researchers and practitioners across countries, in diverse habitat types and with very different socioeconomic conditions. There is clearly still much to be done to improve our understanding of bushmeat–livelihood linkages, particularly at the national level, and how best to address the issue, but this section of the book helps take both of these a step further. As was the message of the conference at large, however, it is clear that situational factors are often determinant. Single response strategies are unlikely to comprehend the diversity of the circumstances in which wild animal populations are found, nor the potential for sustainable off-take from those areas. A more nuanced and situationally specific set of responses is therefore required.

11

Can Wildlife and Agriculture Coexist Outside Protected Areas in Africa? A Hopeful Model and a Case Study in Zambia

Dale M. Lewis

Introduction

Rural poverty and hunger, the prohibitive costs of wildlife law enforcement and human–wildlife conflicts arising from land use practices severely limit wildlife conservation outside Africa's national parks. Developing conservation approaches that are culturally acceptable as well as self-financing in response to these challenges would have enormous implications for maintaining viable large-mammal populations in Africa. In eastern and southern Africa, conservation efforts have placed much emphasis on tourism to link tourism revenues to conservation outside protected areas on community land. Except where human populations are small or revenues exceptionally high, tourism income on a per capita basis is generally too little to modify local land use practices on a scale needed to protect large-mammal species and their habitat.

Zambia provides a valuable 20-year history of research studying wildlife conservation needs outside protected areas (D. Lewis, 2000). What emerges from this work is the critical importance of building strong linkages between agriculture and rural markets with land use practices conducive to improved natural resource management. Recent studies in Luangwa Valley, Zambia, have demonstrated the rationale and methodologies for creating these linkages and how they help reduce important underlying threats to wildlife (D. Lewis and Jackson, 2005; D. Lewis and Tembo, 2000; 2001). These studies further suggest that staying overly focused on wildlife-based markets, which are often constrained by government and private sector control, may impede innovations in wildlife management that could help build more effective synergies for conservation. This chapter brings into focus the results of these studies and their potential relevance to natural resource conservation outside national parks in Africa.

The Zambian situation

Zambia has a protected area system consisting of 18 national parks, where laws prohibit all forms of hunting and human settlements. For the most part, 35 game management areas surround these parks. These semi-protected areas are community lands under customary ownership and, under the Wildlife Act of Zambia, the government grants licences for the legalized hunting of wildlife in these areas and returns a share of these licences to local community authorities called community resources boards. Through these boards the revenues earned from wildlife support community projects as well as wildlife resource protection, but little if any is directed at actual livelihood needs at the level of the household.

Background studies, towards developing a new approach for conservation

Food security and agricultural practices

Food security surveys (D. Lewis, unpublished data) in five game management areas in Luangwa Valley, Zambia, showed that between 20% and 60% of the residents in these areas, who were largely subsistence farmers, failed to grow enough food to last to their next harvest. Poor farming practices, erratic rainfall, inadequate farming inputs and crop damage by wild animals contributed to this variability. In many cases food-impoverished households were forced to provide farm labour in exchange for food or spend time away from their fields catching fish, snaring wildlife or searching for honey as a means to barter for food produced by more successful farmers.

In one of these game managements areas, Mwanya, the WCS (D. Lewis and Tembo, 2000) undertook an intervention to artificially improve food security by providing 120 90-kg bags of maize to 183 households living in Yakobe village, located in a relatively wildlife rich area and where lack of food security was a general household problem. This experiment sought to answer two questions:

- If farmers had sufficient food during their farming season as provided by the intervention, would they then have the means to stay committed to farming and produce a good food crop?
- If these households did achieve increased food security as a result of their more intensive farming efforts, would they be less dependent on wildlife or less prone to illegal hunting to make up for food shortfalls?

Food was provided just prior to the planting season in November 1998. Granaries were resampled in 2000 to compare with levels found in 1998

Table 11.1 Comparison of household food security before (1998) and after (2000) the food security interventions, based on food produced and assessed from household granaries in December.

	Number of households	Population	Total food stock (kg)
1998	183	896	7,940
2000	180	859	34,478

(Table 11.1). Figure 11.1 shows the amount of food per person based on known food stocks in December projected to March when farmers are most active with weeding and guarding their crops. The results showed a relatively food-secure farming season in 2000 compared with 1998.

The potential relationship between these results and the use of snares to obtain game meat was analysed from questionnaire data collected from safari clients who hunted in this same area from 1997 to 2000. Despite efforts from 1997 to 1999 to increase community support for wildlife conservation through wildlife revenue-supported projects and public awareness exercises,[1] data suggested that the use of snares increased during this period (Figure 11.2). In contrast, in the year following the intervention of maize support, when food security levels dramatically improved in Mwanya area, the percentage of clients who complained about snares declined by over 50%.

Markets, household income and income sources

Based on a sample of 1,065 randomly selected households from five different game management areas in Luangwa Valley (D. Lewis, unpublished data), the five most common sources of household income were poultry, rice, ground nuts, beer and fish as ranked against a total of 56 potential income sources. Individuals who owned or operated grinding mills or small shops earned the

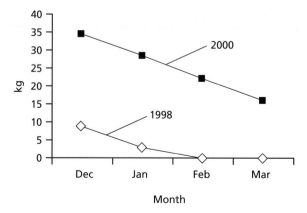

Figure 11.1 **Projected food security levels for pre- and post-food security intervention, showing critical food shortages in 1998 from food locally produced versus a food-secure community in the year following the intervention.**

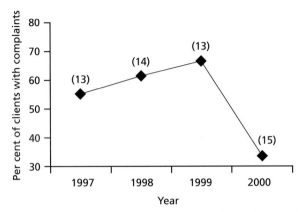

Figure 11.2 **Snaring levels in Mwanya area, based on the incidence of snares encountered by safari hunting clients. Total number of clients (N) sampled for each year shown in chart. Safari hunting data are missing for 2001 and 2002 because of a safari hunting ban.**

highest income, followed by those growing cotton on household plots or in private sector employment. In contrast, poultry farming ranked 34 in terms of income earned, while growing rice, groundnuts, beer and fish ranked 19, 20, 23 and 10 respectively. On average, annual household income was about US$80. Factors contributing to low income were lack of farm inputs, poor market development, low producer price, and inadequate skills and technologies to increase production or diversify income sources. In the absence of alternative trading partners, small-scale traders operating in these areas frequently took advantage of this extreme poverty and succeeded in buying commodities at well below fair prices from families who were too desperate for money to negotiate better prices.

A selected sample of 88 individuals whose livelihood depended largely on illegal hunting with firearms, primarily with locally made muzzle-loading guns, showed that income derived from illegal hunting averaged US$320 with a median of US$161 (Figure 11.3). Income derived from other sources, mostly agriculture, averaged US$170. In many of these cases, illegal hunting income was used to finance farming inputs and farm labour to increase earnings further, suggesting that illegal hunting, as opposed to a more legitimate alternative, was subsidizing agricultural development and increased household income. Hunting was clearly a desirable livelihood option for those owning a gun and having the necessary skills to hunt successfully. From this sample, 32% had been previously arrested, some as many as four times, suggesting that the threat of court convictions was not an effective deterrent.

Wildlife losses

Based on interview data for the above 88 sampled hunters, the annual number of animals killed per hunter averaged 29. Buffalo (*Syncerus caffer*),

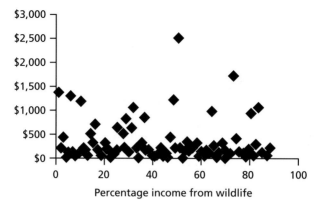

Figure 11.3 **Income distribution for 88 sampled poachers from Luangwa Valley and surrounding areas.**

impala (*Aepyceros melampus*), warthog (*Phacochoerus africanus*) and kudu (*Tragelaphus* spp.) were the four most popular species hunted, related either to their size or relative abundance. Twenty-two per cent of the sample had previously hunted elephants.

Interview data from a sample of 445 local farmers[2] living in areas where wildlife was prevalent, and where many of the above hunters were interviewed, suggested that 57% of the residents in these areas set snares 4.7 times each year and used, on average, six snares on each occasion. This contributed to an average of 5.8 animals snared annually per household, used largely to exchange for food when household granaries ran low and to consume for protein.

Ironically, wildlife in these same areas contributed to improved social needs such as schools and clinics through wildlife revenue-sharing programmes intended to promote increased public support of conservation. Local leadership authorities recognized by the government received these revenues, derived largely from safari hunting, on behalf of the community and financed projects that promoted social needs and services. This contradiction might suggest a lack of alternative, higher-paying markets and the prevalence of chronic food shortages, as documented in the livelihood surveys, thus outweighing household commitment to conservation.

A market-based model to reduce threats to wildlife

These background studies in Luangwa Valley supported the following conclusions:

- Poor subsistence farmers who live in areas with wildlife and who cannot grow enough food to meet their annual food needs are more likely to engage in coping strategies, such as snaring, that threaten wildlife production.

- Agricultural markets that are poorly developed to offer people incentives to become better farmers can contribute to land use practices detrimental to wildlife production. For example, rice and maize were commonly grown in the valley but ranked low as an income source.
- Without external assistance to develop skills and improved community organization, rural communities in general fail to negotiate trade deals favourable to their own livelihood needs or the needs of their resources.
- Wildlife-based tourism revenues that fail to meet basic household livelihood needs cannot sustain household commitment to control land use practices detrimental to wildlife.

These conclusions provided the necessary building blocks to construct an alternative model for wildlife conservation in rural areas surrounding national parks. Referred to as community markets for conservation or COMACO, the model forges practical links between agriculture and rural markets to promote increased community commitment to conservation. Figure 11.4 illustrates the various components of the model, which rely on increased producer prices to encourage farmers to become better farmers, and thus more food secure and more active participants in wildlife conservation. The model links these steps with conservation by requiring farmers to adopt land use practices conducive to wildlife production to be eligible to these increased producer prices.

Key to the COMACO process and the different interventions it supports is a regional trading centre, which operates as a limited company with community ownership but with business management run by qualified external personnel. Called the Conservation Farmer Wildlife Producer Trading Centre (CTC), the trading centre targets, at least initially, households most vulnerable to hunger and poverty and offers them extension services for improving farming skills and improving access to markets, as long as participating households form producer groups and pledge commitment to land use practices conducive to natural resource conservation. By carefully selecting households with livelihood needs that foster destructive land use practices, the CTC can leverage conservation compliance among literally thousands of residents living in areas supporting wildlife. As illustrated in Figure 11.5, the model associates producer groups with producer depots, where goods are bulked and traded through a network of trading hubs linked to the trading centre.

To test the COMACO model, measured change in wildlife numbers was correlated with increased food production and increased producer prices offered to participating households. An important source of help in implementing and testing the COMACO model was the World Food Programme (WFP). Participating households received free maize from WFP as an incentive to learn and adopt improved farming skills and form agricultural producer groups with by-laws stating their commitment to conservation and food security. This included the voluntary surrendering of 15 snares per group. These producer groups, in turn, became the future trading partners of the CTC. Participants

Figure 11.4 **Primary components of the COMACO programme, which targets low-income, food-impoverished households with incentives to become food secure through fair, legal trade benefits that are conditional on compliance with improved land use practices.**

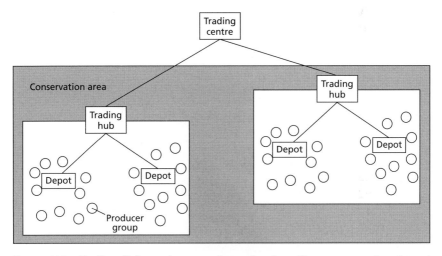

Figure 11.5 **Trading linkages between the regional trading centre and registered producer groups in rural areas supporting wildlife resources.**

received free food for only 1 year as a way to increase their farming effort and gain greater self-reliance in food crop production. WFP support declined as households became food secure. In 2004, for example, annual food relief from WFP in the project area declined by about 40% from the previous year.

Results: COMACO model field testing

Area of operations

Field-testing of the model lasted from 2001 to 2003 and covered approximately 9,000 km^2 of rural landscapes outside Luangwa Valley national parks, where a trading centre in Lundazi and its associated trading depots were established.

This initial COMACO area covered five chiefs' areas, including Mwanya, Chitungulu, Kazembe, Chifunda and Chikwa (Figure 11.6). These are collectively referred to as the COMACO core area. Since 2004, new areas have been added, and these are referred to as the COMACO extended valley area or extended plateau areas.

Food security

COMACO initially targeted, in 2001, about 2,500 households selected as food insecure or unable to grow enough food to complete a farming cycle. By 2004, COMACO had expanded to cover about 25,000 km² with over 16,000 participating households. COMACO interventions achieved significantly improved food security (Table 11.2). Among the participating households, 30–

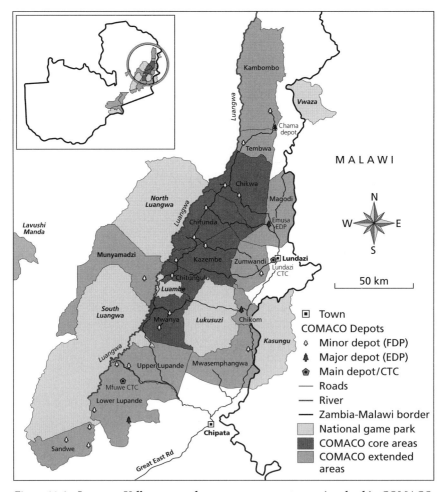

Figure 11.6 **Luangwa Valley area and game management areas involved in COMACO interventions.**

Table 11.2 Compliance and food security results of the 'food for better farming' initiative.

Year	Total farmers per year	Total assessed per year	Total conservation farmers from those assessed	Total h/holds composting from those assessed	% food secure from those assessed	Total farmer groups cumulative
2001–02	2,434	1,584	961	0	30	102
2002–03	5,574	2,697	2,176	1,899	68	371
2003–04	8,621	2,379	1,414	1,373	48	491

Table 11.3 Snares and illegal firearms surrendered by producer group members.

Year	Total snares collected	Total guns collected
2001–02	6,000	95
2002–03	8,752	35
2003–04	15,303	291
Totals	*30,055*	*421*

68% achieved food security their first year based on increased food production from employing improved farming practices, including pot-hole cultivation and composting.

Conservation commitments and land use plans

COMACO's agreement with producer groups was that they comply with their conservation pledges and in exchange they would have access to improved livelihood skills and better-paying markets through the CTC. A summary of the conditions producer groups were asked to comply with are as follows:

- use pot-holing cultivation with compost;
- form farmer groups to encourage more intragroup skills-sharing and confidence-building;
- refrain from use of snares and show this commitment by surrendering snares to the CTC;
- stop burning crop residues after crop harvest; and,
- develop farmer group by-laws that include group commitment to better land use practices and group compliance with community-based decisions that promote natural resource management in their area.

By far the most dramatic demonstration of community compliance to this agreement was the large number of snares and illegal firearms surrendered by producer group members, totalling 30,055 snares and 421 illegal firearms from 2001 to 2003 (Table 11.3). These figures contrasted greatly with parallel efforts by wildlife law enforcement officers, whose approach included patrols in wildlife areas, house searches and road blocks. Total workforce by wildlife police officers in carrying out these operations in the project area averaged about 280,000 man-days per year at an approximate cost of US$2 per man-day (Regional Managers Office Data, 2004). From 2002 to 2003 these law enforcement operations recovered a total of 155 snares and 242 firearms.

COMACO also required producer groups to comply with a land use plan developed by their community leaders through a consultative process with broad community participation to adopt specific steps the community could take to reduce potential conflicts with their natural resources. Compliance with these land use decisions was also a condition to gain access to CTC trade benefits and to have access to the various extension support services provided by the CTC. In 2003, participating communities developed these plans in both map and local language text form and, in 2004, local district council authorities held full council meetings and adopted these land use plans as an approved instrument of local government to promote better land use management. They have become an important basis for community leaders to use market incentives to influence their constituency to resist destructive land use practices and assist in the production of wildlife for added economic benefits through legal markets.

CTC market development, price incentives and crop diversification

Establishment of the CTC and its network of producer depots, trading hubs and associated staff required an initial preparatory phase of infrastructure development, staff training and community orientation and involvement. By 2003, the CTC launched a full season of buying and selling rice, village chickens, honey and groundnuts, which represented commodities the majority of households could produce. The 2003 marketing season provided the necessary information and experience to improve the CTC's capacity to process commodities into packaged, higher-paying products and to expand trade links for improving producer prices.

Since 2003, COMACO has increased producer prices for the various commodities it purchases and processes at levels varying from 58% to 108% (Table 11.4).

Although cotton was a common cash crop in the project area, the CTC did not market or promote this crop because of its adverse effects on soil nutrients and wildlife habitat. To maximize earnings, cotton farmers tended to grow large fields and, because cotton is a labour-intensive crop, farmers with large cotton fields had less time to grow a food crop. Preliminary data from two villages in Chief Kazembe's area, Zokwe and Chabesha, suggested that cotton farmers had less food security than those who did not grow cotton. Generally, after 3 years of growing cotton, farmers who lacked fertilizer were forced to clear new land to sustain high cotton production. Except where rice farming was possible, most farmers complained about the lack of an alternative cash crop. Compared with cotton, it requires only 20% of the effort to grow an equivalent yield of rice, which also has minimal impact on wildlife habitat. For these reasons, COMACO regarded rice as a 'wildlife-friendly' cash crop. To provide a more complete coverage of a second 'wildlife-friendly' cash crop and stimulate economic incentives to promote crop rotation with a legume crop, the CTC introduced soybeans in 2003 and has expanded these efforts for the 2004 planting season.

Wildlife conservation trends: towards establishing a COMACO impact

COMACO attempts to link markets with improvements in the food security of members of rural communities most in need and thus most likely to adopt coping practices that might reduce wildlife numbers and habitat. If these interventions truly diminish potential threats to wildlife resources through reduction of numbers of active illegal hunters or actively used snares, wildlife numbers should show signs of increase or at the very least show less decline than areas where COMACO is not operating. Both of these predictions are examined below.

Aerial transects were flown in 1999, 2002 and 2006 to evaluate population trends for 10 wildlife species for the COMACO core area. Munyamadzi Game

Table 11.4 CTC prices for commodities purchased from COMACO producer groups.

Commodity	Unit	Pre-COMACP price (2002)	2004 COMACO price	2005 COMACO rice	2006 COMACO price
Rice	kg, unpolished	K600 (US$0.1)	K950 (US$0.2)	K1,200 (US$0.3)	K1,000 (US$0.3)
Chicken	One adult size	K5,000 (US$1.1)	K9,000 (US$1.9)	K9,000 (US$2.0)	K9,000 (US$2.8)
Honey	kg	K1,200 (US$0.3)	K2,000 (US$0.4)	K2,500 (US$0.6)	K2,550 (US$0.8)
Groundnuts	kg, shelled	K950 (US$0.2)	K1,200 (US$0.3)	K1,500 (US$0.3)	K1,100 (US$0.3)

Table 11.5 Trends for total COMACO area. *d*-Test comparing 2006 with 1999 and 2002 shown, with significant values in bold ($P < 0.05$, ~ 25 d.f.).

Species	1999*			2002†			2006‡			*d*-Test	
	No. observed	Projected total	SE	No. observed	Projected total	SE	No. observed	Projected total	SE	02	99
Buffalo	1,114	5,542	4,245	746	4,633	3,401	156	1,529	1,211	–	–
Wildebeest	468	2,360	1,476	155	960	583	167	1,651	910	0.64	-0.41
Waterbuck	32	161	100	22	147	80	75	735	483	1.20	1.17
Zebra	114	566	240	93	578	551	116	1,137	788	0.58	0.69
Elephant	296	1,490	701	137	879	398	207	2,055	816	1.30	0.53
Eland	29	144	97	10	58	45	7	69	58	0.15	-0.67
Hartebeest	25	125	81	2	11	10	40	421	162	2.53	1.64
Roan	3	113	12	6	31	26	14	137	133	0.78	0.93
Kudu	22	109	83	14	78	50	13	127	94	0.47	0.14
Puku	391	1,967	1,248	179	979	477	224	2,212	930	1.18	0.16
Impala				678	3,877	1,251	556	5,537	2,296	0.63	–
Warthog				43	223	149	55	540	290	0.97	–
Bushbuck				0	0	0	2	20	18	1.08	–

*Area of transect 885 km², total area sampled 5,307 km², SI 16.67%.
†Area of transect 1,052 km², total area sampled 5,307 km², SI 19.82%.
‡Area of transect 544 km², total area sampled 5,329 km², SI 10.20%.

Management Area was included in the sample because of comparable food security efforts undertaken throughout the same period as the core area, although COMACO market support was not in place in that area. Fixed, parallel aerial transects were flown with sampling intensities varying from 10.2% to 19.8%. Selection of the species censused was based on several criteria, including large body size and vulnerability to snaring because of frequent use of waterholes during the dry season. Use of snares around waterholes is a common practice to take advantage of predictable animal movements. Those species judged to be most at risk to poaching, at both population and individual levels, because of low local density and meeting these criteria included wildebeest (*Connochaetes* spp.), waterbuck (*Kobus ellipsiprymnus*), eland (*Taurotragus oryx*), hartebeest (*Alcelaphus buselaphus*), roan (*Hippotragus equinus*) and kudu. For the purposes of this chapter, we term these the 'poaching-susceptible species'. Elephants were included in our census because of their conservation value. Buffalo were initially included because of their status as the preferred target of illegal gun hunters (data not shown), but accurate estimation of their populations is proving difficult because of their heterogeneous distribution (i.e. 'clumping'), which leads to extremely high variation. Puku (*Kobus vardonii*) were included to represent one of the small-bodied species with a relatively high reproductive rate.

In addition to a species-by-species analysis of the combined area, the study also looked at population trends within each area surveyed to determine if there were local differences in response. Because of their high variance and small sample size within the individual surveyed areas, data for the poaching-susceptible species excepting wildebeest were combined for some analyses. Wildebeest had relatively large group size variation compared with the other species, which generally occurred in smaller groups of 5–12. Therefore, inclusion of this species with the other poaching-susceptible species would have skewed the data.

Tables 11.5 and 11.6 show the combined surveyed area data and the combined population trends, respectively, for the high-risk, low-density species group,. Most species showed positive, upward trends, but these were statistically insignificant. Wildebeest and eland numbers showed negative, downward trends, which again were insignificant. Buffalo were excluded from analysis because of heterogeneity. In the analysis of population trends for individual surveyed areas for the combined poaching-susceptible species, two of the four areas from the COMACO core areas, Chanjuzi and Chikwa, showed significant increases ($P < 0.05$) from 1999 to 2002 and from 2002 to 2006. All the other areas showed positive but statistically insignificant trends, with the exception of Mwanya, which showed a negative but statistically insignificant change.

In summary, population estimates from the 2006 survey allowed a preliminary indication of population change for selected species relative to population sizes prior to COMACO (1999) and at the beginning of the

Table 11.6 Data combined for the poaching-susceptible species, excepting wildebeest. Comparisons between 1999 and 2006, and between 2002 and 2006 are shown, with significant values in bold ($P < 0.05$, ~25 d.f.).

Area	1999			2002			2006			d-Test	
	No. observed	Projected total	SE	No. observed	Projected total	SE	No. observed	Projected total	SE	02	99
Chikwa	14	80	35	24	137	51	53	520	167	**2.19**	**2.58**
Chifunda	3	12	7	11	55	20	14	137	83	0.96	1.50
Chanjuzi	3	17	14	1	6	5	28	303	70	**4.26**	**4.04**
Munyamadzi	45	222	103	17	137	53	42	412	189	1.45	0.88
Mwanya	43	222	80	0	0	0	13	127	72	1.77	-0.88

COMACO intervention (2002). No significant decreases were found for any of the analyses and significant increases were found for hartebeest for the combined surveyed area sample and for two of the four COMACO core areas for the high-risk, low-density species group. Of interest, almost every species showed a downward trend from 1999 to 2002, suggesting that populations were in decline at the beginning of the COMACO programme. Across the board increases in populations would be unlikely within 4 years, given a need to first stabilize these populations, and then allow them to recover.

If this increased wildlife population trend is related to the large number of snares farmers surrendered in compliance with the conditions of the COMACO model, then it is important to assess whether farmers replaced the snares they surrendered and persisted using them as a substitute for farming. Evidence for non-replacement would suggest that the COMACO model had achieved a significant link between better farming practices and improved markets with improved conservation results. A questionnaire completed by safari clients after each safari hunt was the basis for making this determination (Table 11.7). These data were collected in 1999 and between 2003 and 2005. A ban on safari hunting during 2001 and 2002 precluded data collection for these years. Three of the four hunting areas, Chikwa, Chifunda and Mwanya, showed dramatic decreases in snare complaints, with Chikwa area having no complaints throughout the period and Chifunda and Mwanya showing a threefold and twofold decrease respectively. Chanjuzi area showed a reduction and then replacement, indicating that cheating did indeed begin to occur.

Discussion

Poor, unsuccessful farmers living on land outside Luangwa Valley's national parks typically rely on land use practices that contribute to lowered wildlife production. The case study presented in this chapter tested the converse relationship, that food secure farmers who earn income from crops not in direct conflict with wildlife can support land use practices more conducive to wildlife production. The particular model used to test this hypothesis, COMACO, linked market and food security incentives to household compliance with conservation commitments. In terms of its capacity to affect rural communities across large geographic areas, COMACO offers an extremely relevant case study for examining a model that builds synergy between agriculture, rural markets and conservation.

Specifically, the COMACO model influences land use practices outside protected areas by offering better-paying farm-based markets, as well as improved crop production skills, to farmers who support natural resource conservation. The latter is achieved under the guidance and compliance of producer group by-laws and community land use decisions. The process initially targets poor, unskilled farmers because hunger and poverty are the

Table 11.7   **Percentage of snare complaints from safari clients.**

Year	Chikwa	Chanjuzi	Chifunda	Mwanya
1999	15	50	17	62
2003	0	0	0	30
2004	0	83	11	25
2005	0	50	0	27

underlying reasons why many households in Luangwa Valley illegally kill wildlife or unnecessarily degrade wildlife habitat. Key to the model's long-term success is its capacity to self-finance its interventions while attracting an increasing proportion of the community to access trading benefits through a profitably run trading centre in exchange for conservation compliance.

Data presented in this case study suggested that the model is not only viable but is a cost-effective and socially acceptable way to achieve conservation. This was demonstrated by the increase in wildlife production, which correlated with the surrendering of thousands of snares and hundreds of firearms in compliance with conditions for receiving food security support and accessing better-paying markets through the CTC. These data contrasted greatly with the relatively smaller number of snares and firearms recovered from much higher cost law enforcement operations.

Direct maize support from the WFP was an important basis for initiating farmers to learn and practise better farming practices and become future producer groups for the CTC. In addition to providing increased producer prices as a way of encouraging farmers to use their improved farming skills, the CTC also used differential pricing to reward producers who achieved specific conservation targets. For example, producer groups that maintain over 15 beehives receive a higher producer price for honey, and farmers who remain compliant to composting, non-burning of crop residues and conservation farming, while also adopting bee-keeping, receive a higher producer price for rice or soybean.

Based on survey data collected prior to COMACO interventions, it was clear that farming was the most common livelihood, practised by nearly every resident, but with low levels of success for a significant percentage in terms of self-reliance in food production or household income. Farmers had little incentive to modify farming activities in ways more compatible with wildlife production. Rather, for many, wildlife was a convenient way to mitigate hunger or poverty by using game meat as a way of bartering for farm food or supplementing household income. Incentives by the private sector to encourage farmers to grow a cash crop may have exacerbated the problem by giving little consideration to the long-term impact on soil and watershed or to the potential conflict and social consequences caused by certain labour-intensive cash crops. Against this background, studies suggested that poor, food-insecure farmers were taking a heavy toll on wildlife numbers and efforts to contain this problem by conventional law enforcement measures were largely ineffective. It is thus argued that the economic incentives and services provided by the CTC under the COMACO model reduces the need for higher-cost interventions to police natural resources, respond to famine needs, or maintain costly rural welfare programmes.

During its relatively short period of interventions, COMACO has had a profound impact on snare and illegal firearm removal from large areas surrounding some of Zambia's most prestigious national parks and has provided incentives for households to adopt farming practices more conducive to both wildlife production and increased food security. COMACO's economic sustainability remains untested, however. The viability of the model depends on the self-financing capacity of the CTC to provide marketing and livelihood support services indefinitely. It also requires that the total commercial benefits returned to local communities under COMACO and through the efforts of the CTC be greater than commercial benefits provided by competing markets. Otherwise, low-income household commitment to their conservation pledges will probably diminish as well as their loyalty to producing commodities for the economic viability of the CTC. This was a problem in 2004 when the CTC encountered difficulties in maintaining payment schedules to producer groups and households changed their loyalty to other traders, even though prices were lower than those offered by the CTC. As a business, COMACO, through its regional CTCs, has the potential of sustaining improved household income and monitoring of this impact relative to other traders is an ensuing part of the COMACO programme.

Again, the model under test has shown real potential for influencing land use practices and provides improved prices to a growing number of registered producer groups. Given the free-market environment in Zambia, however, the CTC has a major challenge in remaining solvent and competitive, especially as much of its profit goes towards sustaining farmer commitment to food self-reliance and conservation. If it were to fail because of these environmental and social service objectives, wildlife-based revenues from tourism would gradually decline, resource protection costs would increase, land management would suffer and the cost and need for government-level programmes to support farmers would rise. For these reasons, the Zambian government not only has a stake in COMACO's success but could also play a decisive role in its future by helping integrate COMACO into a national strategy for rural development and natural resource management.

Recommendations

Having demonstrated COMACO's initial success, the following recommendations could help guide the Zambian government's efforts to expand and sustain COMACO's support services for rural communities which share their lands with key natural resources of national importance:

1   The exemplary role by WFP in using food relief as a means to promote improved farming practices and to reduce specific environmental threats for which selected households may be responsible should be a more widespread policy in Zambia. If it were, such interventions could be combined with efforts to establish farmer groups, which in turn would facilitate the establishment of trading centres with their network of producer depots in other regions of the country.

2   Designate the current project area a pilot National Conservation Area to more fully test the COMACO model in terms of its economic sustainability, rural livelihood benefits and conservation impact across a broad ecological landscape. Such a designation would facilitate enhanced inter-ministry support of the COMACO model and would promote efforts to refine the model's applications to natural resource management and rural livelihoods elsewhere in Zambia.

3   Based on the above economic feasibility and impact studies, the government should be prepared to offer a relatively low-cost subsidy to the CTC to minimize government's own rural development support costs while gaining a much greater range of investment returns related to rural livelihoods, tourism development and natural resource conservation.

4   Commercial cash crops known to have adverse affects on soils and watershed needs, such as cotton and tobacco, warrant an environmental impact study before such crops are promoted on a large scale or allowed in areas with high watershed, wildlife or tourism potential.

Conclusion

COMACO, as presented in this chapter, is not a finished product that can stand on its own. It remains dependent on support from grants and outside assistance. It does represent a concept under development in a landscape characterized by many of the land use conflicts currently degrading natural resources throughout much of Africa. It also represents one of the largest scale experiments to test the relationship between improved livelihoods and natural resource management. The approach is also not one that relies on a 'passive' cause-and-effect relationship, but rather takes a more proactive approach by forging linkages through well-developed market incentives directed primarily at households who are most vulnerable and thus most likely to degrade natural resources.

COMACO is relatively new, though it has derived from a long history of lessons from previous Community Based Natural Resource Management (CBNRM) programmes in Zambia, most notably the ADMADE programme. Past programmes such as these, as well as the adaptive management of COMACO in the first 3-year pilot phase, enabled COMACO to be developed initially, and potentially replicated elsewhere. Results from this pilot initiative provide growing evidence that rural development synergies with conservation are possible and may be sustainable if the business model built around regional trading centres proves viable. Such assurances are not possible from the results presented here but the potential benefits to wildlife conservation were demonstrated.

While COMACO has received grants and financial support to launch its concept as a rural development approach to conservation, it is not a typical donor project. More a business model, COMACO has received these grants much as any large-scale business venture would leverage investments to derive profit and increased production. What is atypical of COMACO, however, is that its commercial production comes from producers who were targeted for their lack of production skills. This added to the cost of initiating COMACO, but as the process builds improved human capacity and local partnership to supplying CTCs with their products, the cost of production, in theory, will decrease and conditions for sustainability will improve. This is the stage the COMACO model has currently reached and the next phase of the COMACO experiment will be to this critical assumption of COMACO's sustainability.

Acknowledgements

The WCS acknowledges the Royal Norwegian Embassy, the World Food Programme, the Wallace Global Foundation, the Japanese Embassy and Care International for their support in COMACO's development. A USAID SANREM/CRSP project in collaboration with Cornell University has assisted WCS in the analysis of COMACO results. Particular gratitude is extended to the Zambia Wildlife Authority for its partnership, friendship and collaboration with WCS, without which COMACO's development would not have been possible. Similar gratitude and appreciation is extended to the local government authorities of Lundazi, Chama, Mambwe and Luangwa Districts, to the community resources boards and to the traditional leaders of these districts for their enthusiastic support in promoting COMACO in their areas, and to the Ministry of Tourism, Environment and Natural Resources for its growing interest in COMACO's contribution to conservation in Zambia. The WCS also acknowledges the dedication and years of hard work by its Zambian staff, who have worked tirelessly to build COMACO from the bottom up. Their names are too many to cite, but special recognition is given to Nemiah Tembo, James Phiri, Whytson Daka, Jolum Jere, Ruth Nabuyanda, Mike Matokwani, Malambo Maambo, Kabila Mukanda, Bennet Siachoona and Handsen Mseteka.

Notes

1 The interventions were part of a community-based wildlife management programme called ADMADE that returned safari revenues to local communities whose areas produced wildlife utilized by safari hunting. Income received sustained the costs of village scouts to guard against poaching and to support community projects such as schools and clinics. ADMADE-employed staff also facilitated community training to improve public understanding for conservation and legal use of wildlife.

2 The WFP and WCS collected these data in 2003 from a sample of approximately 4,000 households which had participated in a food security initiative supported by these organizations to encourage the voluntary removal of snares. After 2 years of community participation in this programme during which time no-one was arrested for surrendering snares, the interview was administered to ask respondents about their prior history of setting snares and assumed that answers were honest as there was perceived to be no threat of being arrested.

12

Food for Thought for the Bushmeat Trade: Lessons from the Commercialization of Plant Non-timber Forest Products

Elaine Marshall, Kathrin Schreckenberg, Adrian Newton, Dirk Willem te Velde, Jonathan Rushton, Fabrice Edouard, Catarina Illsley and Eric Arancibia

Introduction

During the past 20 years, the commercialization of non-timber forest products (NTFPs) has been widely promoted as a means to achieving the sustainable development of tropical forest resources (de Beer and McDermott, 1989; Nepstad and Schwartzman, 1992; Arnold and Ruiz Pérez, 1998). This interest is based on earlier perceptions that forest exploitation for NTFPs can be more benign than for timber (Myers, 1988), together with a growing recognition of the subsistence and income generation functions of NTFPs in the livelihoods of the rural poor, and especially women (Wollenberg and Ingles, 1998; Neumann and Hirsch, 2000; Angelsen and Wunder, 2003; Ruiz Pérez *et al.*, 2004). Following the 'use it or lose it' philosophy, it was hoped that NTFPs could be valuable enough to outcompete other land uses, such as agriculture, ranching and logging, which were significant threats to forest cover. Much of the NTFP literature over this period has focused solely on plant NTFPs – and it is in this sense that we will use the term – with a parallel body of literature dealing with issues around the use of bushmeat.

The early NTFP literature focused on the conservation benefits that could be derived from the use of NTFPs. More recently, in line with the growing global interest in rural poverty reduction, the main focus has been on the potential development benefits of NTFPs. Here too, there has been an evolution in the debate, beginning with a strong focus on recording the subsistence role of NTFPs to one more concerned with NTFPs as sources of income that could provide a route out of poverty. Most recently, there has been a particular interest in the commercialization of NTFPs and in understanding the circumstances under which it can successfully meet the dual objectives of improved livelihoods for the rural poor and sustainable management of forest resources (Neumann and Hirsch, 2000).

Within this context, the Forestry Research Programme (FRP) of the UK Department for International Development (DFID) funded a multidisciplinary research project entitled Commercialisation of Non-timber Forest Products in Mexico and Bolivia: Factors Influencing Success (CEPFOR). The CEPFOR project undertook participatory fieldwork with four NGO partners in Mexico and Bolivia, focusing on 10 different plant NTFPs (Table 12.1) in 18 communities. In addition to research at community and household level, the project took a value chain approach, describing the full range of activities required to bring each product from producer to consumer and emphasizing the value that is realized in the process. A more detailed presentation of the project's methods and results is provided in Marshall *et al.* (2006).

In this chapter, we present some of our key findings structured around a series of questions that are frequently asked in the NTFP literature. In the concluding section, we reflect on how each of these questions may also be relevant to the bushmeat debate.

What constitutes successful commercialization?

There has been very little discussion about what constitutes success with respect to NTFP commercialization (Marshall *et al.*, 2003), yet an understanding of how people define success under different circumstances is a prerequisite for the development of appropriate policy interventions. Traders, for example, may be interested primarily in profit margins, whereas producers also cite

Table 12.1 **CEPFOR case study NTFP species and their uses.**

Country	Product	Species name	Part used
Bolivia	Organic cocoa	*Theobroma cacao*	Beans
	Natural rubber	*Hevea brasiliensis*	Latex
	Incense and copal	*Clusia* and *Protium* spp.	Resin
	Jipi-japa palm	*Carludovica palmata*	Leaf fibre for weaving
Mexico	Soyate palm	*Brahea dulcis*	Leaf fibre for weaving
	Maguey	*Agave cupreata*	Plant heart fermented to produce alcohol
	Mushrooms	*Boletus edulis*	Fruiting body fresh and dried
		Tricholoma magnivelare	
		Amanita caesarea	
		Cantharellus cibarius	
	Pita	*Aechmea magdalenae*	Leaf fibre for embroidery
	Camedora palm	*Chamaedorea elegans*	Fresh leaves as floral greens
	Tepejilote	*Chamaedorea tepejilote*	Inflorescence as food

compatibility with other livelihood activities, or involvement of rural women in income-generating activities. Broadly speaking, value chains can be successful in a number of different ways (Schreckenberg *et al.*, 2006):

- volumes or values traded via different routes and incomes generated – both overall as a contribution to local and national economies and for the individual actors concerned;
- governance of the chain – the rules governing the relationships (preferably transparent) between different actors and the sharing of benefits (preferably equitable) between them;
- sustainability of the chain, or the ability to deliver a consistent supply to meet demand over the long term – incorporating not just social and economic, but also environmental, sustainability;
- achievement of a range of locally defined objectives.

Working with local communities, traders, development NGOs and government officials, the CEPFOR project identified many different definitions of success, ranging from economic (household income generation) to social (the activity does not cause internal conflict in the community) and environmental (the rate of commercialization is consistent with biological sustainability). Objectives differ at household level (the activity helps meet the family's basic needs, the work is agreeable), at community level (a high proportion of community members benefit, the activity increases community prestige) and at national level (increased tax revenues, improved consumer welfare). Definitions of success are also dynamic, changing in response to variation in socioeconomic circumstances and the behaviour of the market. Effective project interventions therefore require discussion with communities and other stakeholders in the NTFP value chain, to jointly identify criteria of success, evaluate the trade-offs that might be needed between them, and establish systems for monitoring progress.

Who is involved in NTFP commercialization?

Is NTFP commercialization the preserve of the poorest?

The CEPFOR project was specifically interested in measuring the involvement of the poorest in NTFP trade. Well-being ranking was used to identify the poorest in each case study community. As shown in Table 12.2, this revealed that NTFP trade is not solely the domain of the 'poorest of the poor'. Nevertheless, in Bolivia, where some of the case study communities are extremely remote, over one-third of NTFP households surveyed depend on NTFP trade for more than half their annual income which, in 70% of the communities, is less than US$1 per capita per day. Furthermore, the NTFP trade is frequently their only

Table 12.2 **Relationship between well-being level and engagement in NTFP activities (modified from Marshall *et al.*, 2006).**

Well-being ranking of NTFP households relative to non-NTFP households	% of communities ($n = 13$)
NTFP households are disproportionately concentrated in lowest well-being group	39
NTFP households are disproportionately concentrated in middle well-being group	15
NTFP households are disproportionately concentrated in top well-being group	15
Not applicable as all households involved in NTFP activity	31

source of cash income. This is less likely to be the case in Mexico, where the case study communities have easier access to other income-generating activities, as well as agricultural subsidies and various forms of social security, resulting in per capita incomes that are generally between US$1 and US$3 per day (only two communities having less than US$1 per day). Only 15% of Mexican NTFP households relied on NTFP trade for more than half their annual income.

Although the level of benefits derived from NTFP commercialization depends on many factors, such as the level of involvement of the individual, household size and life cycle stage, and ability to be combined with other activities (see below), the NTFP trade enabled many interviewees to accumulate substantial savings. These were invested in a variety of activities such as re-roofing their homes and covering their health costs.

Is NTFP trade particularly beneficial to women?

NTFP activities are considered to be particularly attractive to women because they typically involve small-scale activities that require few skills and low capital investment and combine well with traditional domestic roles (Falconer, 1996; Fontana *et al.*, 1998). In none of the product value chains observed in the CEPFOR project are women solely responsible for all activities. Many activities are economically viable at household level only because they involve both men and women, making best use of their skills and opportunity cost of labour (Marshall *et al.*, 2006). Whereas men may have alternative income-generating activities available to them (such as working as hired labour or in the mines), NTFP activities often represent one of the few cash-generating opportunities for women in marginalized rural communities. This means that women will engage in NTFP activities even if they generate very low returns to labour, whereas men may choose to pursue more lucrative activities. In one Bolivian community, for example, men collect rubber for several weeks at a time, achieving returns to labour of US$4.20/day, which compare favourably with the local daily wage rate of US$3.40. As processing the rubber has a return rate only of US$2.50/day, however, this task is left to women while the men migrate to work in the nearby mines, returning only once their wives have finished processing.

Women are more likely to be involved in an NTFP activity if it is close to home and are more likely than men to be involved in processing and cultivation activities (Figure 12.1). Harvesting of products at some distance from the community is predominantly the role of men, except in the case of products that can be either sold or consumed at home, such as mushrooms. Financial benefits tend to be greatest for women if they are involved in selling and can decide about how the income is used. Even when they do not generate a large absolute income for women, we found that NTFP activities can provide women with self-confidence and improved status. This is particularly likely if the work is undertaken as part of a group, when social networking, training and group savings activities are also important benefits for women. This is illustrated by the case of a jipi-japa weavers' association established by a private company near Santa Cruz, Bolivia, to ensure supply of high-quality artefacts for its tourist shops at the same time as supporting local women. Members are trained to produce a range of decorative items for which they receive a guaranteed price (from US$1 to US$12). A small additional payment is made by the shop into a social fund (to cover women's health costs) and into a rotating fund, which the women take turns to use for investments such as re-roofing their house.

Who else is active in the value chain?

Much of the NTFP literature focuses predominantly on producers. This reflects the difficulties of accessing other actors in value chains, particularly where these are narrow with just one or two intermediaries at every stage in

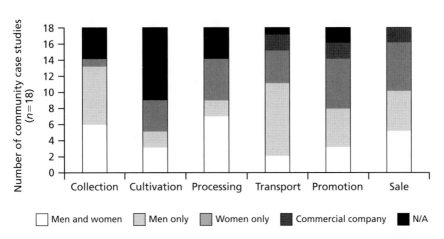

Figure 12.1 **Involvement of men, women and/or a commercial company in different stages of the value chain (from Marshall et al., 2006).**

the chain, or illegal. Nevertheless, it is important to be aware of other actors in the value chain and the power that they may exert over the way in which the trade operates. In addition to producers, who may be harvesters of the wild product or cultivators of a semi-domesticated NTFP and may operate either individually or in an organized group, the CEPFOR project identified a wide range of actors in NTFP value chains including community-based traders, traders from outside the community, community-owned enterprises, cooperatives, associations, government departments and NGOs. Only in the shortest chains are all activities from production to sale carried out by the same individual. In all other chains, each actor is responsible for a specific set of activities including provision of labour, technical expertise, marketing skills, financial support, etc. Concentration of power (defined as the degree to which a single individual or firm controls the value chain) was found to be greatest in value chains producing highly processed or perishable products for a sophisticated international market. This is a result of the distance to market information sources and the low education levels of producers. Understanding the specific role of each actor in the value chain is essential if conditions for the poorer actors in the chain are to be improved.

Do NTFP activities contribute to poverty reduction?

The importance of NTFPs in sustaining people's livelihoods is widely accepted, with much evidence documenting the fact that forest-dependent people often have few options except to gather and hunt NTFPs for their food, medicines and cash income (Falconer, 1990; 1996; Scoones *et al.*, 1992; FAO, 1995; Ros-Tonen, 1999). This is one of the main drivers behind donor support to NTFP commercialization initiatives. However, an increased focus of development policy on poverty reduction has brought with it a need for unequivocal and more differentiated evidence about whether and how much NTFP commercialization can contribute to poverty reduction (Wunder, 2001; Arnold, 2002).

NTFP harvesters are typically people who live at the margins of economic and political systems (Shanley *et al.*, 2002). NTFP activities are considered to be attractive to such resource-poor people, despite the fact that they are characteristically labour intensive, because they generally have low technical entry requirements, can provide instant cash in times of need and the resource is often freely accessible (Neumann and Hirsch, 2000). Paradoxically, it has been suggested that the same characteristics that make NTFP activities attractive to poor people also make them economically inferior activities (Angelsen and Wunder, 2003). Not only do they yield low returns and offer little prospect for accumulating sufficient capital assets to escape poverty (Ashley *et al.*, 2003), but the arduous nature of the work may mean that people will not engage in them if there are alternatives, they may be vulnerable to substitution by cheaper synthetic or industrial alternatives, and ease of entry may lead to excessive competition and inability to generate a surplus from production and sale. This has led to NTFP activities being labelled a potential '*poverty trap*' that keeps people in chronic poverty.

None of the CEPFOR cases fell into this category. However, they could be grouped into three types of NTFP activity with respect to poverty reduction (Marshall *et al.*, 2006):

- '*Safety nets*' prevent people from falling into greater poverty by reducing their vulnerability to risk. They are particularly important in times of crisis and unusual need, with many families engaging in NTFP activities only when subsistence agriculture or cash crops fail, or when illness hits the family. It is particularly true of products that are available all year round because, as one Bolivian incense collector explained, 'the knowledge that incense is available to be harvested and traded acts as a guarantee that, no matter what, some income can be earned'.

- Most of the CEPFOR case study products fall into the '*gap-filling*' category, in which NTFP-derived income is supplementary to more important farm and off-farm income-generating activities. These activities are carried out on a regular basis, often in the non-agricultural season, and contribute between 7% and 95% of cash income to the household. The proportion of income contributed depends in part on the other economic opportunities available to families and on the seasonal availability of the product, with those that are available for longer periods of time often contributing more to the household economy (Figure 12.2). The CEPFOR study confirmed that, although many NTFP-based activities generate only small amounts of income, the timing of this income during the non-agricultural season may increase its relative importance (e.g. Alexiades and Shanley, 2004). They play a key role in income-spreading and generally make poverty more bearable through improved nutrition or higher income but do not necessarily make people less poor. However, as shown in Figure 12.3, even products that contribute only a relatively small share of a household's cash income may be perceived to play a valuable role in a household's livelihood strategy.

- '*Stepping stone*' activities help to make people less poor. Ruiz Pérez *et al.* (2004) suggest that it is only in areas that are well integrated into the cash economy that some NTFP producers are able to pursue a 'specialized' strategy in which the NTFP contributes more than 50% of total household income and collectors and producers tend to be better off than their peers. In the CEPFOR cases, no single NTFP activity could be classed as a stepping stone for all participants. Nevertheless, depending on the degree of intensity with which a household engaged in them, several activities can act as stepping stones for individual families. In the case of wild mushroom harvesting in Oaxaca, for example, occasional collectors only collect about 10 kg per season compared with 70 kg for average collectors and 300 kg for a small

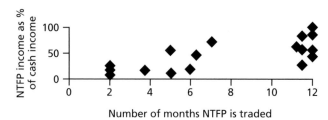

Figure 12.2 **Relationship between months traded and the proportion of cash income provided by NTFPs in the case study communities (from Marshall *et al.*, 2006).**

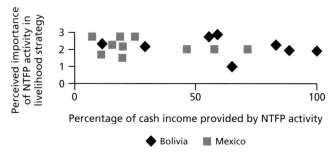

◆ Bolivia ■ Mexico

Figure 12.3 **Importance of NTFP income in livelihoods in case study communities (from Marshall *et al.*, 2006).**

group of serious collectors (Marshall *et al.*, 2006). Depending on where they sell the mushrooms, the last group can achieve an income of up to US$475 during the short mushroom season, sufficient to enable them to send a family member to work in the USA.

Are NTFP activities viable on their own?

An interesting finding of the CEPFOR project is that many NTFP activities appear to be successful because of their association with other activities. Examples include:

- *Incense collection.* Incense is harvested from relatively rare tree stands that are 2–4 days' walk from the case study community in Beni, Bolivia, requiring significant investment to organize collecting trips. The risk of not finding incense is counteracted by the possibility of also collecting copal, a less valuable but more common resin than incense, and bushmeat, along the way.
- *Soyate palm processing.* The leaves of this palm are plaited and sold on for processing into hats and other woven products. Although the returns to labour are negative, all members of the case study community in Guerrero, Mexico, plait soyate leaves as it can be carried out alongside other activities – anywhere and at any time, except in church.

- *Organic cocoa trade.* In one Bolivian community, accessible only by river, organic cocoa beans and dried fish are bought by the same traders, with cocoa dominating from December to February, when dried fish is not available.
- *Fresh mushrooms.* Near Oaxaca, Mexico, fresh mushrooms, which are available for only 2 months, are taken to market by a community-based trader who travels to town every week to sell the bread that she makes and other farm produce that she buys from households in the community.

This suggests that, at all levels, whether collection, processing or sale, NTFP activities frequently benefit from being associated with other activities. The cocoa case illustrates the flip side of this complementarity, however, with the cocoa trade being both dependent on the dried fish trade and highly vulnerable to any changes that may occur in it. This implies that efforts to intervene in the NTFP trade must also take into account the likely impact on the associated activity.

What impact does NTFP commercialization have on the natural resource base?

Early interest in NTFPs was encouraged by the belief that NTFP commercialization that added sufficient value to forest products could contribute to forest conservation (Nepstad and Schwartzman, 1992). Since then, research has suggested that the conservation benefits of NTFP commercialization may be somewhat limited, and there is growing evidence of unsustainable harvesting of NTFPs leading to the degradation of forest resources (Boot and Gullison, 1995; C. M. Peters, 1996; Wollenberg and Ingles, 1998). This is confirmed by the CEPFOR case studies, in all of which harvesting NTFPs for trade initially led to resource overexploitation. This is only the beginning of the story, however, as the communities have found a number of ways to reduce the negative impacts on the resource. These include displacement activities such as harvesting from further afield or buying in the resource from other production areas, but they also include improved management of the wild resource and domestication. It is also important to note that, in several of the CEPFOR cases, land use conversion is a much greater threat to the resource than NTFP commercialization.

Management of the wild resource

In 35% of the cases, communities are trying to enhance the management of the wild resource and improve harvesting practices. This is particularly true in the case of communally owned resources, where communities are engaging in land use zonation, rotation of harvesting areas, training of harvesters and monitoring of the impact of harvesting on the resource. Although often initiated by communities, such efforts have frequently benefited from external support to provide ecological knowledge and organizational capacity.

Domestication

When land is held privately, and the plant can be easily propagated, individuals often begin to engage in small-scale domestication, as is true of 45% of the CEPFOR cases. Domestication generally consists of transplanting wild germplasm to establish a resource closer to home, rather than improving its quality. External support is important to successful domestication, as are appropriate biological characteristics of species, the socioeconomic and legal context of domestication and the level of traditional knowledge about the product. Whereas community management of the resource is more likely to continue to benefit the original collectors, the move towards domestication is nearly always disadvantageous to the poorest because of their lack of financial capital to cover establishment costs and, in some cases, because of their lack of land. Furthermore, there is a real risk of industrial plantations displacing both harvesters who collect wild resources and small-scale cultivators.

What is the role of entrepreneurs in commercialization?

Entrepreneurs or intermediaries are undoubtedly the most maligned actors in NTFP value chains, frequently accused of exploiting producers. Yet, one or more entrepreneurs were key to establishing or sustaining almost every one of the CEPFOR value chains (te Velde *et al.*, 2006). They can play a particularly important role in long value chains where products are exported outside the country.

Typically entrepreneurs are private individuals acting as intermediary traders, but they may also be presidents of producer associations or NGO staff members. Their essential roles are to achieve one or more of the following (Marshall *et al.*, 2006):

- Bridge information gaps. Lack of market information (on price, quality and quantity) is the key barrier for poor people to enter the NTFP trade. Many producers have little idea of the ultimate destination of the products they sell.
- Identify new market niches. This role is best played by people who have a foot in both the producer and the consumer environment. Thus, export of fresh matsutake mushrooms to Japan was initiated by a Mexican of Japanese descent, while community members who have moved to the capital, La Paz, are the key go-betweens in the Bolivian incense trade.

- Provide training and information to ensure product quality, and to help organize communities. Some entrepreneurs have social motives and wish to help particular groups of people in addition to achieving commercial success.
- Help gain physical access to markets.
- Advance capital to ensure consistent product supply. Access to credit is almost non-existent for most rural NTFP producers so traders may also provide credit to cover the costs of inputs.

Notwithstanding their positive role, there is always a concern that intermediaries may exploit their position of power, leading – in the worst case – to debt peonage, as in the well-documented case of wild rubber harvesters in the Brazilian Amazon (Schwartzman, 1992). This was not evident in any of the CEPFOR cases, although in certain value chains a large proportion of the benefits do appear to remain with the entrepreneurs, together with most of the decision-making power. As shown in Figure 12.4, such concentration of power is more likely in longer value chains. Some justification for this is provided by the financial risks involved in these value chains, e.g. in the form of cash advances required to secure refrigerated transport for fresh mushrooms or camedora palm fronds. A way of reducing such inequitable distribution of benefits is therefore to reduce the risk along the value chain. Rather than by-passing intermediaries, evidence from the CEPFOR project indicates that it may be better to work *with* them in order to provide them with a more reliable and consistent supply line and, in return, to increase the proportion of benefits received by the community. At the same time, efforts to improve market information at community level are essential to increase the ability of communities to negotiate prices and define the rules of trade. The CEPFOR project found that this kind of bargaining power is vital in determining the satisfaction levels of poor producers, processors and traders in NTFP value chains.

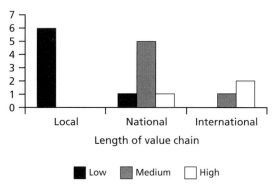

Figure 12.4　**Length of value chain and concentration of power Note: concentration of power is defined as the degree to which a single individual or firm controls the value chain; scored subjectively on the basis of market reports (from Marshall *et al.*, 2006).**

Does successful commercialization require organization?

A key factor determining success of the CEPFOR cases is the ability to innovate. In some cases, innovation was introduced to the value chain by an entrepreneur or external organization. However, in many cases, innovation – whether in terms of resource management or market strategies – was initiated by a community-level organization. Community-level organizations exist in 60% of the CEPFOR cases, though fewer than 15% have achieved real horizontal integration, in which all producers (or processors) work together. NTFP activities are often taken on by pre-existing organizations in a community (e.g. water committees, women's groups, church groups). The existence of traditional knowledge of a product (and the resource), as well as of marketing other agricultural products, are important factors in determining a community's interest and capacity to successfully commercialize NTFPs. Community organizations dealing with NTFPs often follow a similar path:

- Initial focus on ensuring product quantity and quality, as insufficient or inconsistent quantity and quality of product almost inevitably results in traders ceasing to visit a community.
- Depending on the constraints, focus on resource management, processing and/or arranging for more cost-effective transport.
- Finally, work on improving relationships (or negotiating power) with traders and consolidating market niches.

Many community-based producer organizations also provide additional benefits by contributing funds to community projects.

Horizontal integration is not the only model that works for NTFP commercialization. Vertical integration – in which successive stages in the value chain are placed under the control of one enterprise – has been found to be important in facilitating the establishment of specialized niche market value chains. This was particularly evident in several entrepreneur-dominated chains exporting products out of the country. As discussed in the previous section, this can lead to control of profits by the lead individual (the person who controls the market information and the capital), but it does not necessarily lead to exploitation of producers.

How important is legality?

One of the sources of risk in NTFP and bushmeat value chains alike is illegality. In Bolivia, as in many other countries, there is very little policy or legislation specific to NTFPs. In Mexico, there is overregulation, with NTFPs mentioned in a great many different laws. Both scenarios lead to confusion about which laws apply and who is responsible for implementing them. Therefore, producers in the informal sector, where much NTFP trade takes place, rarely consider the legislative context to be a great constraint. Nevertheless, communities are often obliged to remain in the informal sector because they lack the capacity to fulfil the legal requirements for formal sector NTFP commercialization. In Mexico, for example, harvesting permits for some products, such as wild mushrooms, which are on the national protected species list, require environmental impact assessments (EIAs) that are beyond the technical or financial reach of small-scale producers. Not only do they cost about US$10,000 to carry out, but their remit is narrowly confined to the plant's ecology rather than providing information that might help communities develop sustainable management plans. One solution for making them more accessible to communities would be to allow for a single EIA to be undertaken for a group of neighbouring communities harvesting the same product.

By enabling communities to operate in the formal sector, legality can bring benefits of increased security for producers, processors and traders. In the Chinantla region of Mexico, for example, SEMARNAT (the Mexican Ministry of the Environment and Natural Resources) has been encouraging producers of camedora palm fronds to obtain harvesting permits. This would enable them to sign agreements with buyers, guaranteeing prices and medical assistance (collection of the fronds is dangerous) in return for a specified volume and quality of palm fronds. In spite of technical support from SEMARNAT and financial help available from the Forest Development Programme (PRODEFOR), the requirements are so complex that most producers continue to harvest palm leaves without the permits specified by the law.

Another benefit of legality is that it is a basic requirement for participation in certification schemes, e.g. for organic or fair trade products, which may provide access to new and more profitable markets. The costs of achieving certification, like legality, may be beyond the reach of individual producer communities. Intervention by NGOs or others is likely to be needed to help carry out ecological impact studies, develop management plans, monitor the resource and the division of benefits, etc. Where possible, such support needs to be coordinated with relevant government programmes, such as grants or subsidies for specific permit or certification requirements.

Conclusions

In this paper we have presented some lessons that are generalized from a group of 18 product/community case studies in NTFP commercialization in Mexico and Bolivia. Although we recognize that bushmeat cases, like NTFPs, are extremely varied and need to be examined on a case-by-case basis, we hope that some of the following issues may be of interest within the bushmeat debate.

- *Defining success.* Based on the experience of the CEPFOR project, we suggest an ideal definition of successful NTFP commercialization as a transparent, equitable and sustainable activity that has a positive impact on poverty reduction, gender equality and resource access, tenure and management (Marshall *et al.*, 2006). It is very important, however, to review objectives on a case-by-case basis and with all the different actors involved in a particular value chain. Bushmeat hunters, for example, are likely to have very different objectives from women sellers of bushmeat stew or end consumers and, only if all are taken into account, can interventions in the value chain succeed.
- *Who is involved?* Linked to the previous point, it is important to understand all the actors involved in the bushmeat trade. Value chain analysis is a useful tool to help identify all the actors and their costs and benefits, and to determine who exerts power and takes different decisions.
- *How much does bushmeat contribute to poverty alleviation?* If poverty alleviation is a concern, it is necessary to understand the specific role bushmeat plays in sustaining local livelihoods: does it act only as a safety net, is it a regular gap-filler or can it even provide a few people with the stepping stone they need to escape poverty altogether? Successful interventions need to target these specific functions of bushmeat, e.g. by providing alternative safety net activities, making gap-filler activities more widely accessible or increasing the proportion of money realized from stepping stone activities that stays in the community.
- *Are bushmeat activities viable on their own?* Before intervening in a bushmeat value chain, it is essential to understand the dynamics of any other activities with which it may be associated.
- *What is the impact on the resource?* As in the case of plant NTFPs, it is likely that tenure of bushmeat species (or the environment in which they live) will play an important role in determining whether the response to any overexploitation is better resource management and/or domestication. The CEPFOR experience suggests that domestication would not necessarily benefit the same people as are initially engaged in bushmeat exploitation.
- *Importance of entrepreneurs.* Many value chains are established and sustained by entrepreneurs. Before intervening to bypass them, it is important to understand their positive contributions that might be made to the value chain, as well as the power they exert over other actors in the chain.
- *Importance of organization.* Community-level organization is essential for management of communal resources, as is often the case for bushmeat species, and can strengthen the ability of producers and processors to negotiate with intermediaries.
- *Legality.* This is a major discussion point in the bushmeat literature. Lessons from the NTFP trade suggest that legality can increase security for producers, reduce justification for intermediaries to retain 'unfair' shares of the profit in the value chain and open the doors to certification. At the same time, a

legal trade is usually heavily regulated and difficult to access, which may introduce new expenses for a community.

The flora and fauna NTFP literatures have been developing in parallel for some years now. We hope that this chapter can contribute to an exchange of learning and experiences between them with the aim of developing rigorous but adaptable research frameworks, effective enabling policies and targeted interventions to protect and support small-scale producers and traders in meeting their livelihood needs in a sustainable manner.

Acknowledgements

This publication is an output from a research project (Project R7925 Forestry Research Programme) funded by the UK Department for International Development (DFID) for the benefit of developing countries. The views expressed are not necessarily those of DFID. The results presented in this chapter are drawn from a joint publication (Marshall *et al.*, 2006), which documents a research project coordinated and implemented together with the CEPFOR team. We gratefully acknowledge the input of all these colleagues to the work on which this chapter is based.

13

Bushmeat, Forestry and Livelihoods: Exploring the Coverage in Poverty Reduction Strategy Papers

Neil Bird and Chris Dickson

The relevance of the poverty reduction strategy process to forests and bushmeat

Since 1999, poverty reduction strategies (PRSs), as documented in poverty reduction strategy papers (PRSPs), have become the national development framework in many countries. This World Bank and International Monetary Fund (IMF) initiative has now been endorsed by other multilateral and bilateral donors, and forms the framework for much international development assistance (World Bank and IMF, 2002). PRSPs are intended to be country-owned documents, implying the leadership of national governments and the involvement of civil society (Booth, 2003). They are designed to improve the comprehensiveness of poverty reduction measures over past efforts in an effort to achieve the millennium development goals. Over 50 developing countries have prepared PRSPs (World Bank, 2004). Originally set as a requirement for debt relief under the enhanced heavily indebted poor countries (HIPCs) initiative, many non-HIPCs have also invested in preparing these plans. Additionally, access to the poverty reduction and growth facility (PRGF) of the IMF is now conditional on a nationally owned PRSP. Key policy measures and structural reforms aimed at poverty reduction and growth are identified and prioritized during the PRSP process, and, if feasible, their budgetary costs are assessed. The PRSP is thus a formal representation of a nation's development policies, and helps determine the attitude of the international community towards national efforts. Although not in itself a guarantee of funding, inclusion in the PRSP is a necessary platform to gain political prominence. Whether an issue is included in these documents consequently has a bearing on the likelihood of implementation and success of any given initiative.

Within the broad coverage of environmental issues, sustainable management of forests has been the subject of considerable sectoral analysis over the last decade. By adapting the volume and method of off-take of timber and non-timber forest resources, it is hoped that the interests of poverty reduction and resource conservation can be reconciled. However, the issue of bushmeat shows discrete socioeconomic and ecological characteristics, and merits attention separately from more general forestry issues. In particular, the mobility of the resource and its fugitive nature (in the sense of not being owned until the point of capture and death) leads to difficulties in measurement and regulation of the 'stock'. Equally, the informal nature of much of the trade makes it difficult to assess or formalize trading activities. Yet such an analysis is important, as much of this economic activity is believed to be carried out by members of poorer communities. Within these groups, the financial benefits often constitute a large proportion of household income, as well as being an important source of protein. Under these conditions, bushmeat resources are an indispensable safety net for those most vulnerable to environmental or seasonal fluctuations in resource availability (de Merode *et al.*, 2003). There is therefore a clear link between bushmeat activities and poverty.

This chapter explores the presentation of bushmeat issues within PRSPs, including relevant references to biodiversity conservation or more general forest policies, in an effort to see what prominence these issues are currently given.

Coverage of forestry and bushmeat in PRSPs

Selection of countries

The countries chosen for this study are known range states where bushmeat is a significant economic activity. Although frequently seen as essentially a West and Central African phenomenon, similar activities are common throughout the world, albeit often under different names.[1] In an effort to understand the wider picture of how bushmeat issues have been treated in PRSPs, this chapter includes a number of Central and South American and South-East Asian countries, in addition to the West and Central African states. In all these countries the forest sector is a significant part of the economy. The sample therefore allows for a comparison in the treatment between bushmeat and other forest resources that may contribute to improved livelihoods of the poor. Table 13.1 lists the countries and provides some forestry and poverty statistics. Poverty levels are high throughout, with approximately 40% of most populations living below the locally defined national poverty line.

Search methodology

The method used in this assessment was adapted from two similar studies that have examined the inclusion of environmental issues (Bojö and Reddy, 2002) and forest issues in PRSPs (Oksanen and Mersmann, 2003). For 11 countries within the sample, both the interim and final PRSPs were reviewed.[4] For the remaining five countries the final PRSP document has yet to be published, in which case the country analysis depended on a review of the interim PRSP. Automatic word searches were carried out on each PRSP (or I-PRSP) for a

Table 13.1 Countries reviewed in this study with selected forestry and national poverty indicators

Country	Country area (km²)[2]	Forest area as % of total land area (2000)[3]	Population 2002 (millions)[3]	Percentage of population below the national poverty line[3]	Percentage of population below US$1 a day poverty line (2002)[3]
DR of Congo	2,344,860	60	52	–	–
Indonesia	1,904,570	58	212	27	8
Bolivia	1,098,580	49	9	63	14
Tanzania	945,090	44	35	36	49 (1990)
Nigeria	923,768	15	133	–	70
Zambia	752,612	42	10	35	64
Central African Republic (CAR)	622,984	37	4	–	67 (1990)
Cameroon	475,442	51	16	40	17
Vietnam	331,690	30	80	42	18
Cote D'Ivoire	322,463	22	17	–	16
Uganda	241,040	21	25	44	85
Ghana	238,538	28	20	40	45
Nicaragua	129,494	27	5	48	50
Benin	112,622	24	7	33	–
Honduras	112,090	48	7	–	21
Sierra Leone	71,740	15	5	–	–

–, no data available.

number of forestry and bushmeat-related terms. The terms used were as follows:

Forestry terms	Bushmeat terms
forestry	bushmeat/wild meat/game meat
forest resources	wildlife trade
forest management	wildlife products
tree products	hunting/trapping/trophy hunting
non-timber forest products	community wildlife management

Management of renewable natural resources

Instances of these phrases in each of the documents were extracted and compiled in two tables. Each mention was then evaluated according to the following criteria in order to assess the degree to which the issues described had been incorporated in the PRS process:

(i)   'Issue assessed': forest or bushmeat issues are mentioned in the poverty assessment/analysis.
(ii)   'Linked to poverty': causal linkages between forest or bushmeat-related issues and poverty-related issues are discussed in the documents.
(iii) 'Responses discussed': forest or bushmeat-related responses and actions are defined in the documents.
(iv)   'Processes discussed': process links between the PRSP process and forest or bushmeat-related policy and planning processes are described in the documents.

Within each of these categories, the treatment of the issue was given a score out of 3, where:

0 = not mentioned
1 = mentioned but not elaborated
2 = elaborated
3 = best practice

Where does bushmeat appear and how is it treated?

Quantitative overview

Forestry issues are mentioned in 20 out of the 28 documents reviewed, and in all of the full PRSPs. In contrast, bushmeat is mentioned in only seven of the documents, and never scores at a level higher than level 1 (mentioned but not elaborated) (Table 13.2). Although some of the documents (e.g. from Bolivia, Zambia, Uganda) mention policies and initiatives to maximize sustainable

Table 13.2 Scoring of bushmeat and forestry issues according to the Bojö and Reddy method. Scores in italics indicate those allocated by Oksanen and Mersmann (2003).

Country	Document*	Forestry				Bushmeat			
		Issues assessed	Linked to poverty	Responses mentioned	Processes discussed	Issues assessed	Linked to poverty	Responses mentioned	Processes discussed
Benin	I-PRSP, 2000	0	0	0	0	0	0	0	0
	PRSP, 2003	1	1	2	2	0	0	0	0
Cameroon	I-PRSP, 2000	0	0	2	1	0	0	0	0
	PRSP, 2003	2	1	2	2	0	0	1	1
Central African Republic	I-PRSP, 2000	1	2	2	1	1	0	1	0
Cote D'Ivoire	I-PRSP, 2002	2	0	2	0	0	0	0	0
DR of Congo	I-PRSP, 2002	1	0	1	0	1	1	0	0
Ghana	I-PRSP, 2000	0	0	2	2	0	0	0	0
	PRSP, 2003	1	1	2	2	0	0	1	0
Nigeria	CAS, 2002	0	0	0	0	0	0	0	0
	NEEDS, 2004	1	1	0	0	0	0	0	0
Sierra Leone	I-PRSP, 2001	1	1	1	0	0	0	0	0

Country	Document								
Zambia	I-PRSP, 2000	0	0	0	0	0	0	0	0
	PRSP, 2002	1	1	2	1	1	1	1	1
Tanzania	I-PRSP, 2000	0	0	0	0	0	0	0	0
	PRSP, 2000	1	0	0	0	0	0	0	0
	Draft PRSP, 2004	1	1	1	0	0	0	0	0
Uganda	I-PRSP, 2000	0	0	0	0	0	0	0	0
	PRSP, APR 2003	2	2	1	1	1	0	1	0
Bolivia	I-PRSP, 2000	0	0	0	0	0	0	0	0
	PRSP, 2001	1	1	2	1	1	0	1	0
Indonesia	I-PRSP, 2003	0	0	0	0	0	0	0	0
Vietnam	I-PRSP, 2001	0	0	0	0	0	0	0	0
	PRSP, 2003	1	2	2	1	1	0	0	0
Honduras	I-PRSP, 2000	1	2	1	1	0	0	0	0
	PRSP, 2001	2	2	3	2	0	0	0	0
Nicaragua	I-PRSP, 2000	1	0	1	0	0	0	0	0
	PRSP, 2001	1	0	2	1	0	0	0	0
Total		**22**	**18**	**31**	**17**	**4**	**1**	**6**	**2**

*1, not mentioned; 2, mentioned but not elaborated; 3, elaborated; 4, best practice.

Abbreviations: PRSP, poverty reduction strategy paper; I-PRSP, interim poverty reduction strategy paper; CAS, country assistance strategy (precursory document to the PRSP, essentially similar); NEEDS, national economic empowerment and development strategy (effectively the Nigerian PRSP); APR, annual progress report.

NB: although JSAs (joint staff assessments) and PSIAs (poverty and social impact assessments) were read for a selection of these countries, these documents offered no coverage of bushmeat-specific issues.

exploitation of wildlife (ecotourism, ranching, restricted hunting for export), they do not show how these initiatives might relate to current bushmeat off-take. This reflects a more general pattern in the coverage of forestry and bushmeat. There are many suggested policy responses, but fewer references to assessment or analyses of the problems. For example, the Zambian PRSP lists deforestation as the fifth of five problems imposing the greatest social costs upon the Zambian people, but does not explain what the consequences are and how these impact welfare or poverty. However, the suggested response includes extensive details of how to substitute charcoal fuel use and stimulate ecotourism. In addition, both policy initiatives and contextual assessments appear far more often than the two other aspects covered, namely the causal links with poverty and with sector policy and planning processes.

Preservation or exploitation?

Quite a few countries (Bolivia, Benin, Cameroon, Ghana and Vietnam) recognize the importance for local economic development of attempting to stabilize bushmeat practices within sustainable off-take levels. However, the policy recommendations seldom address the bushmeat issue explicitly and independently, often grouping all forestry resources together or grouping them with agricultural interests. A small number of PRSPs take a more utilitarian view, in which the aim is to exploit the economic potential of their wildlife for the whole country. This approach is focused more on the longer-term benefits of wildlife conservation, including ecotourism initiatives, the export of all NTFPs and the growth of agro-forestry and sustainable (often community-based) forestry management. However, there is little focused attention on specific mechanisms to incorporate wildlife into future local livelihoods.

Comparison with forestry content: asset or constraint?

Table 13.2 clearly shows that forestry receives considerably more attention across the PRSP process than bushmeat. Furthermore, in the case of forestry, attention is focused on its positive value as an asset, whereas the position of bushmeat is ambiguous. Twenty two documents consider forestry resources a productive resource, whilst in only eight documents is bushmeat-related activity seen in this way (Table 13.3). On the other hand, 12 documents cite forestry concerns as a constraint, as against five which hold the same view on bushmeat. It is worth noting that the underlying situation and requirements are very different in each of the countries, and the options differ accordingly in the various regions. West Africa is on the whole characterized by high human population, long-standing agricultural activities and associated bushmeat trade. On the other hand, the Congo basin is still a forest frontier, with a sparser human population and a high density of wild game. This game includes many rare and forest-dependent species. The woodland habitat is different again, with an abundance of terrestrial mammals. Differences in the treatment of bushmeat/wildlife in PRSPs partially reflect these obstacles and opportunities in determining sensible and specific steps forward on the bushmeat issue.

Effect of legality on wildlife hunting and policy

The PRSPs do not contain much discussion on the legality of the bushmeat trade. However, it is widely acknowledged that such regulations that do exist are often a legacy of colonial jurisdictions that were not designed to stimulate entrepreneurial, sustainable, decentralized economic activity of any kind. Furthermore, even in those cases where there are regulations in place which might promote sustainable off-take levels and harness the bushmeat trade for local poverty reduction, enforcement capacities are weak. In Ghana, where bushmeat hunting is allowed, this activity is acknowledged within the PRSP, whereas there is no mention of it in the Côte d'Ivoire's PRSP, where such hunting is banned. This is not surprising, as it would be very difficult for a country to make bushmeat hunting illegal but also explicitly support its use for the purposes of poverty reduction. However, hunting of bushmeat species still occurs.

From a conservation perspective, if stricter regulations lead to less wildlife hunting, this might imply that one should advocate a more comprehensive ban on the practice. However, this brief comparison also suggests that a ban on hunting makes it more difficult for bushmeat issues to be legitimately included in the PRS. This in turn deprives wildlife management programmes of the political capital and donor funding that inclusion in the PRSPs may facilitate.[5] The questions that remain are therefore: What are the conservation implications of 'declassifying' game hunting, from a comprehensive ban to more of a 'controlled use' status? And what are the conservation benefits if wildlife use consequently becomes more prominent within the PRSP framework?

The implications of value accumulation along the bushmeat supply chain

The dynamics of the bushmeat supply chain receive little attention in the PRSPs reviewed, for several reasons. First, there are many cultural factors that probably cause under-representation of the true volume of trade. Its taboo or illegal nature does not encourage those involved to divulge their activities. Second, its informality means that much of the trade does not involve money and is therefore difficult to quantify in terms of the value chain. And, third, the supply chain is often both short and localized. Although one might think that this should make it easier to identify and assess, in fact this means that the data regarding revenue are hard to collect and difficult to extrapolate from one area to the next. If there were more exports and the trade more formal, then tax revenues and other indicators would provide a better indication of the distribution of the benefits. However, the short supply chain does mean that

Table 13.3 **Assessment of the treatment of bushmeat as an asset or a constraint in PRSPs.**

Country	Document	Constraint	Process or outcome?	Asset	Process or outcome?	Independently
Benin	I-PRSP, 2000	No	–	No	–	No
	PRSP, 2003	No	–	No	–	No
Cameroon	I-PRSP, 2000	No	–	No	–	No
	PRSP, 2003	**Yes**	**Both**	No	–	**Yes**
Central African Republic	I-PRSP, 2000	No	–	**Yes**	Process	No
Cote D'Ivoire	I-PRSP, 2002	No	–	No	–	No
DR of Congo	I-PRSP, 2002	**Yes**	Neither	Yes	Neither	No
Ghana	I-PRSP, 2000	No	–	Yes	Process	No
	PRSP, 2003	**Yes**	Process	**Yes**	Process	**Yes**
Nigeria	CAS, 2002	No	–	No	–	No
	NEEDS, 2004	No	–	**Yes**	Neither	No
Sierra Leone	I-PRSP, 2001	No	–	No	–	No
Zambia	I-PRSP, 2000	No	–	No	–	No
	PRSP, 2002	No	–	**Yes**	**Both**	No
Tanzania	I-PRSP, 2000	No	–	No	–	No
	Draft PRSP, 2004	No	–	No	–	No
Uganda	PRSP, 2000	No	–	No	–	No
	PRSP APR, 2003	No	–	**Yes**	**Both**	**Yes**
Bolivia	I-PRSP, 2000	No	–	No	–	No
	PRSP, 2001	**Yes**	**Both**	**Yes**	**Both**	**Yes**
Indonesia	I-PRSP, 2003	No	–	No	–	No
Vietnam	I-PRSP, 2001	No	–	No	–	No
	PRSP, 2003	**Yes**	**Both**	No	–	No
Honduras	I-PRSP, 2000	No	–	No	–	No
	PRSP, 2001	No	–	No	–	No
Nicaragua	I-PRSP, 2000	No	–	No	–	No
	PRSP, 2001	No	–	No	–	No

the benefits of improved resource management can be directed towards the poor without being diluted or diverted by many different agents (Cowlishaw et al., 2004).

From a more poverty-centric perspective (such as that of the PRSPs), the potential for economic growth from this kind of initiative is limited. Equally, collecting the data to understand the bushmeat supply chain in one area or community is time-consuming and expensive, relative to gathering other data relevant to poverty reduction. And finally, once produced, the findings from one community would probably not provide 'best practice' guidelines, reproducible at a national level. These factors together may legitimately prevent PRSPs from considering the supply chain of the bushmeat trade in any detail.

Together with other forestry resources	As part of rural development	As part of agriculture and/or fisheries	As part of manufacturing or export resources	For tourism potential	Clash of agricultural and forestry interests
No	No	No	No	No	No
No	No	No	No	No	No
No	No	No	No	No	No
Yes	Yes	No	No	Yes	Yes
No	No	Yes	No	No	No
No	No	No	No	No	No
Yes	No	No	No	No	No
Yes	Yes	No	No	No	Yes
Yes	Yes	Yes	Yes	No	Yes
No	No	No	No	No	No
No	Yes	No	No	No	No
No	No	No	No	No	No
No	No	No	No	No	No
No	Yes	Yes	No	Yes	No
No	No	No	No	No	No
No	No	No	No	Yes	No
No	No	No	No	No	No
No	Yes	No	No	Yes	No
No	No	No	No	No	No
Yes	No	Yes	Yes	No	No
No	No	No	No	No	No
No	No	No	No	No	No
Yes	No	Yes	No	No	No
No	No	No	No	No	No
No	No	No	No	Yes	No
No	No	No	No	No	No
No	No	No	No	No	No

Potentially relevant aspects *not* covered in PRSPs

Management and measurement

There is almost no discussion of what might be 'sustainable' in the context of bushmeat production. In a narrow sense, 'sustainable' off-take can be defined as a rate of depletion that does not exceed the rate of regeneration (natural or assisted). This principle, of limiting off-take to the rate of regeneration, could perhaps be usefully applied to bushmeat activity. However, there are several problems associated with this. First, the stock of mobile fauna is difficult to identify, and is not delimited by national geographical boundaries. This makes it difficult both to determine populations and 'replenishment' rates and to enforce access and property rights. The validity of 'substitutions' is equally difficult – restocking wildlife is harder than reforestation. Substituting with ranched livestock is also problematic due to the unsuitability of species, or the socioeconomic changes that this would impose on the affected communities (Bowen-Jones *et al.*, 2002).

Conflicts

There are very few concrete policies suggesting how to manage bushmeat resources.[6] Policies that are in place usually involve designating areas where hunting is restricted, presumably to allow for the affected species to live undisturbed. In other sections of the PRSPs, however, there are concrete outcome-based recommendations (often relating to infrastructure or industrial development) that potentially clash with such abstract conservationist commitments. Many PRSPs see a combination of structural reform and economic growth as the twin driving forces of poverty reduction. However, road developments cut across swathes of forest and the improved access to remote areas is likely to stimulate bushmeat trade along with other economic activity. This conflict is not addressed in policy recommendations.[7]

Undeveloped linkages

Some of the policies explored in other contexts might facilitate the sustainable utilization of bushmeat resources. Potential benefits include protecting those resources from potential extinction and increasing income security for those dependent on them. However, such linkages remain largely unexploited within the PRSPs reviewed in this study. For example, many PRSPs suggest institutional reforms to improve land tenure conditions, or property rights for the poor. This might increase the sustainability of the bushmeat trade, as formalizing the income might assist regulation against the use of poison or automatic weapons in hunting. Furthermore, secure tenure or usufruct rights might encourage hunters to moderate their off-take in the interests of ensuring future stock levels. Equally, legitimization of this economic activity could help ensure that any reform of the bushmeat market (for example allowing limited export) would benefit the poorest, rather than traders. However, reforms are usually focused on agricultural and sometimes forest exploitation, and seldom mention hunters' rights.

Participatory assessments

Many PRSPs make use of participatory poverty assessments in determining the 'grass-roots' causes and manifestations of poverty. This is a measure to address the possible disconnect between increased nominal income and levels of welfare. When consulted, the rural poor tend to cite irrigation, deforestation, soil degradation, etc., as their major environmental concerns. Depletion of bushmeat stocks does not register as a concern through this method of data collection. At the national level of the PRSP therefore, participation does not seem to reveal a concern for sustainable wildlife management. It is possible that these are simply not voiced by affected communities, or that they are 'filtered out' because they are not shared by all the poor across the nation. There are also concerns about the discourse of the data collection methodology, which may discourage the poor from listing secondary or non-monetary income, and predispose them towards prioritizing growth opportunities rather than vulnerability mitigation (Brocklesby and Hinshelwood, 2001).

Possible causes of low coverage

Low visibility

Consumptive use of wildlife seems to be frequently viewed as affecting only populations within tightly defined geographical or socioeconomic categories.[8] As these are often relatively 'voiceless' groups, their interests may be underrepresented in the PRSPs. This underlines the fact that the contribution of wild meat to local livelihoods is rarely valued in these documents, despite the growing evidence that it may contribute significantly to the well-being of the poor.

Low impact

It is widely recognized that bushmeat is an important source of income or an essential safety net for many of the world's poorest people. However, it is still hard to demonstrate the positive impact on those people of any given policy designed for the sustainable management of wildlife extraction for consumptive use. Furthermore, once those benefits have been quantified, they need to be evaluated in the context of national poverty reduction in order to be relevant to the PRSP. If the benefits of a bushmeat policy are difficult to predict or measure, or if they accrue to only a small proportion of the nation's poor, then that policy will receive correspondingly little attention in the PRSP.

A controversial issue

Over issues such as education or health care, there is usually at least a buildable consensus between the interests of donors, local policy-makers and their constituents. However, this is much harder in the case of bushmeat and forestry resources. The existence value of certain bushmeat species or forest habitats can mean little to those who depend on those resources for income or food. Conversely, international conservationists can have an incomplete understanding of the local dynamics of forest use, which can lead to inappropriate measures being proposed to achieve the conservation of these resources. Although PRSPs are intended to be country-owned documents and not subject to such considerations, they are not immune from donor priorities and sensitivities (Wilks and Lefrançois, 2002).

Within the culture of the international donor community, there are two very different rationales, both of which resist the inclusion of forestry and bushmeat resource management in PRSPs. First, the conservationist perspective stigmatizes the utilitarianism exploitation of wildlife, and certain natural habitats (e.g. tropical rain forests). And, second, there is a potentially counterproductive difference in priorities within the international donor community. As the PRS process gains momentum among donor organizations, particular interest groups are anxious to ensure that their area of interest is included in the format. This includes organizations keen to improve the sustainability of natural resource management in developing countries. However, the PRS process places a strong emphasis on economic growth and infrastructure development. It is therefore difficult for conservationist initiatives to gain the poverty reduction credentials necessary to secure a legitimate and effective place within a poverty reduction strategy.

Bushmeat as a forestry anomaly

The institutional reforms that are often advocated to improve the management of other forest resources do not transfer very well to bushmeat issues. The reasons for this include the informal nature of the bushmeat trade and the fact that bushmeat, by definition, is more mobile than other forestry resources. The use of land is therefore less intense but requires more forest to be set aside. This also means that bushmeat species may roam across borders from one jurisdiction to another, either out of the 'designated' area for controlled hunting, or even into another country with a completely different regime in place. Together, this may in part explain the omission of bushmeat from policies dealing with land tenure, property rights, forestry resources and food security. Although the relevant regulatory frameworks might be well suited to the management of wild game, they seem to apply only to the land itself, not to the animals living on it.

Institutional weaknesses

In many of the countries examined, the design and implementation of sustainable, pro-poor forestry policy is further hampered by certain characteristics of the institutions currently in place. First, the benefits of conservation do not accrue immediately or directly to those who exercise restraint or comply with prohibition. In other words, the preservation of future stock levels is not in itself reward enough for a hunter to reduce his off-take. This is due to a variety of institutional factors, including culture (the bounty of the forest, the substitutability of different game species[9]) and property rights. As people do not feel adequately compensated for the loss of potential game meat, incentives exist for the enforcement agencies to adopt more of a 'rent-taking' attitude towards their conservationist duties (Bowen-Jones *et al.*, 2002). Furthermore, in many countries the institutions underlying natural resource regulation date from colonial times, and are not conducive to the development of sustainable small-scale, decentralized economic activity. This is compounded by the corresponding shortage of skilled and motivated manpower, which not only makes policy formulation difficult, but also impedes the effective collection and analysis of relevant data. These institutional factors make it difficult to establish and enforce sustainable, pro-poor forestry policies, especially in the face of political resistance to measures that may restrict off-take levels in the short term.

Conclusions

An important consequence of the low coverage of bushmeat in PRSPs is that this issue is unlikely to appear high on the national political agenda, which is much influenced by the poverty reduction debate. As a result, not only is there little incentive for coordination across government – an aspect much needed when dealing with natural resources – but also limited support can be expected to be forthcoming from international donors. This will tend to maintain, or worsen, the existing national funding crisis for conservation.

The issue of bushmeat does not benefit from any concrete, outcome-based policy recommendations in these documents. This is not for lack of quantitative analysis as there has been plenty of work in this area, including that carried out by the WCS, the Durrell Wildlife Conservation Trust and the Institute of Zoology.[10] It may be possible to mainstream this research into the World Bank poverty reduction machinery, as (1) much of it focuses on the socioeconomic aspects of the bushmeat trade;[11] (2) the World Bank processes recognize the importance of localized poverty impacts through poverty and social impact assessments, and take account of non-monetary aspects of poverty through participatory poverty assessments (PPAs); and (3) the guidelines for these mechanisms specify that pre-existing relevant research should be used where possible.

However, a number of obstacles remain. Lack of clarity over sustainable best practice in bushmeat issues is a major issue. Allied to this, the growth potential of the extraction of bushmeat from wild populations seems limited. Given the growth emphasis of the PRSs, it is unlikely in any event that an initiative to support *sustainable levels* of off-take would be a valued contribution to a PRS. Even if such potential were identified for some species, there remains the stigma of bushmeat trade as a productive activity in the eyes of some within the international conservation community. These factors combined would make for a lot of work in order to increase the coverage of the issue in PRSPs.

Furthermore, the nominal inclusion in PRSPs is no guarantee of additional support or funding. This relies on the presence and nature of the indicators, the underlying institutional capacity and the follow-up financial mechanisms within government [e.g. the medium-term expenditure framework (MTEF) and state budgets]. The question remains therefore, whether the interests of sustainable wildlife management would be best served by working towards greater legitimacy within poverty reduction, or by increasing efforts through other national and local conservation programmes.

Acknowledgements

This chapter is an output from a research project (ZF0201 Forestry Research Programme) partly funded by the UK Department for International Development (DFID) for the benefit of developing countries. The views expressed are not necessarily those of DFID. Support was also forthcoming from the Zoological Society of London and the MacArthur Foundation.

Notes

1 For the purposes of this chapter, the term 'bushmeat' is used to refer to any meat killed for sale or consumption that was not raised domestically.
2 FAO product and trade statistics, available at: http://www.fao.org/forestry/.
3 Statistics on international development, 2004 edition. DFID, UK.
4 Including Nigeria's CAS and NEEDS documents.
5 Although the linked between PRSP inclusion, funding and implementation are not completely clear and warrant further analysis.
6 This may in part be due to a lack of consensus between local beneficiaries and international donors regarding the trade-off between environmental and economic benefits.
7 For example, Ghana PRSP 2003, point 4.1, plans to commit to three major new highways but makes no mention of the potential impact on natural resources.
8 See the Bolivia PRSP 2001, point 106.
9 Glyn Davies, presentation to the ODI/ZSL Bushmeat and Livelihoods conference, 23–24 September 2004, http://www.odi-bushmeat.org/conference_overview.html
10 See http://www.odi-bushmeat.org/ and http://www.zoo.cam.ac.uk/ioz/projects/ bushmeat.htm for details, also Elliott (2002).
11 For example, http://www.odi-bushmeat.org/download_files/wpb7.pdf.

14

The Beverly and Qamanirjuaq Caribou Management Board (BQCMB): Blending Knowledge, People and Practice for Barren-ground Caribou Conservation in Northern Canada

Ross C. Thompson

Introduction

The Beverly and Qamanirjuaq herds of barren-ground caribou (*Rangifer tarandus groenlandicus*) are two of six major herds of barren-ground caribou in northern Canada. Named after the lakes where they give birth to their calves, these herds are nomadic and range in northern Manitoba, northern Saskatchewan, the Northwest Territories and Nunavut (Figure 14.1). Both herds spend from April to October on the tundra before moving to or past the tree line between November to March, to winter in more sheltered areas. Today, the Beverly and Qamanirjuaq caribou herds are healthy and bountiful. Surveys conducted in the summer of 1994 estimated the Qamanirjuaq herd at 496,000 (plus or minus 105,400 according to confidence levels) and the Beverly herd at 276,000 (plus or minus 111,000 according to confidence levels). These are high densities in historical terms.

The situation has not always been so encouraging, however. This chapter traces the part that new and more participative arrangements have played in improving the management of the herds, and in securing their long-term viability. Whatever the differences between caribou management in the peri-Arctic tundra and the management of bushmeat in the heartlands of tropical Africa, it is suggested that the underlying management message may be the same, straightforward as it is. It is that involving the users of the resource in its management, and doing so in ways that give those users real authority, are important conditions for success. Conversely, marginalizing the resource users, or giving them merely tokenistic authority, are almost sure-fire ways to disincentive the users, with effects that are all too easy to predict.

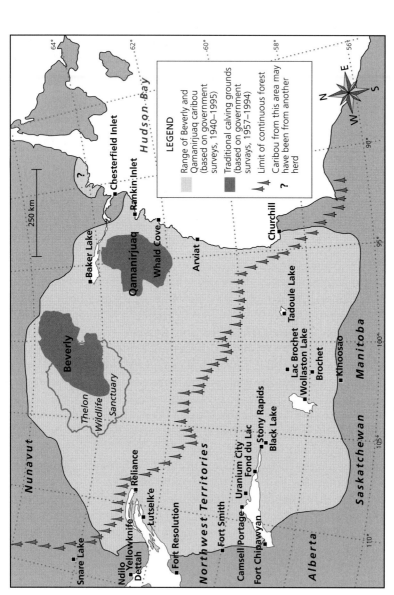

Figure 14.1 Map of Beverly and Qamanirjuaq caribou range.

The Caribou 'crisis'

In the late 1970s, population estimates created concern that the Qamanirjuaq herd faced possible extinction within a decade. This herd was estimated at 39,000 in 1980 – down from more than 145,000 in the early 1950s. This appeared to be a serious decline. The Beverly herd mirrored the downward spiral. A 1980 estimate of 105,000 animals was down from 177,000 in the previous 1974 census.

To respond to the perceived crisis, a caribou management technical group representing government agencies was established in late 1979. Aboriginal users of the herds refused to accept population data in light of their own traditional knowledge and experiences on the land. They claimed that the animals had merely moved to another area, and that government surveys were wrong. Governments recognized that changes in management strategy were needed, but were unwilling to accept the insistence of aboriginal people that their traditional knowledge and practices gave them the right to manage the caribou in their own way.

In the winter of 1979–80, the Beverly herd wintered in northern Saskatchewan in areas accessible by road, and 15,000 to 20,000 animals were killed. Dead animals appeared to have been abandoned on frozen lakes with only their hindquarters and/or tongues removed. Photographs and articles published in the national and international press embarrassed federal, provincial and territorial governments. Aboriginal people were blamed for this slaughter and for the decline of caribou numbers. Provincial governments said that Indians were a federal responsibility and that Ottawa should do something to control them. Ottawa took the position that caribou, being game, were a provincial/ territorial responsibility, and that it was up to those governments to control the hunting of caribou. A confrontational atmosphere developed between native people and wildlife managers.

In December 1980, federal, provincial and territorial ministers met in Winnipeg, Manitoba, to discuss the developing crisis. They agreed that cooperative action was needed and that a solution to the problem had to fully involve aboriginal people. After a series of consultations, there was agreement to form a management board, the Beverly–Qamanirjuaq Caribou Management Board (BQCMB), in which aboriginal people held the majority. Eight individuals would represent aboriginal people, and five individuals would represent the governments of Canada, Manitoba, Saskatchewan and the Northwest Territories (Box 14.1).[1]

The Beverly–Qamanirjuaq Caribou Management Agreement

The original agreement was signed in 1982 between governments of the Canadian provinces of Manitoba and Saskatchewan and the Northwest

---

Box 14.1 The people of the caribou

The Dene. Etthen-eldeli-dene are known as the 'caribou eaters'. Before the arrival of the fur trade, Dene were nomadic, following caribou north to the summering grounds on the tundra and returning to forested areas in the winter. The nomadic lifestyle persisted until the 1940' and 1950s, when the Dene began living in settlements strategically located on major caribou migration routes.

The Caribou Inuit: Ahialmiut. These inland Inuit were given the name 'Caribou Eskimo' by Europeans because the Ahialmiut subsisted almost entirely on caribou year-round, unlike other Inuit, who are mainly coastal and harvest sea animals.

The Cree and Metis people also harvest barren-ground caribou, especially when the caribou migrate deep into Manitoba, Saskatchewan, southern portions of the Northwest Territories and Alberta.

Communities on the caribou range include Fort Smith, Fort Resolution and Lutselk'e in the Northwest Territories; Baker Lake, Chesterfield Inlet, Rankin Inlet, Whale Cove and Arviat in Nunavut Territory; Camsell Portage, Uranium City, Black Lake, Stony Rapids, Fond du Lac and Wollaston Lake in Saskatchewan; Brochet, Lac Brochet, Tadoule Lake, South Indian Lake and Churchill in Manitoba; and Fort Chipewyan in Alberta.

About 13,250 aboriginal people live on or near the range of the two herds. The total estimated harvest in 2000–01 was 18,500 animals.

Territories and Canada. It was reconfirmed in 2002 until 2012, the parties being the governments of Canada, Manitoba, Saskatchewan, Northwest Territories and Nunavut territory.

The agreement commits the parties to fund and support the BQCMB. The parties shall annually contribute CA\$15,000 each (CA\$1 approximately equivalent to US\$0.88). Also, each party is responsible for funding expenses of its appointed board members, including community representatives. In turn, the board is obliged to:

- develop and make caribou and habitat conservation and management recommendations to governments and traditional users;
- monitor habitat;
- conduct information programmes;
- assess and report on herd management plans;
- submit annual reports;
- develop rules and procedures for its operations.

The board consists of 14 members, including a chairman and vice-chairman. Nine individuals represent aboriginal people, and there is one representative each from the governments of Manitoba, Saskatchewan, the Northwest Territories, Nunavut and Canada. There are:

- two Inuit members from communities in the southern Kivalliq (Keewatin) region of Nunavut;
- one Dene member from a community in the South Slave region of the Northwest Territories;
- one Métis member from a community in the South Slave region of the Northwest Territories;
- three Dene members from communities in northern Saskatchewan;
- two Dene members from communities in northern Manitoba;
- a senior official from Indian and Northern Affairs Canada;
- a senior official from Manitoba Conservation, government of Manitoba;
- a senior official from Saskatchewan Environment, Government of Saskatchewan;
- a senior official from Resources, Wildlife and Economic Development, government of Northwest Territories;
- a senior official from the Department of Sustainable Development, Nunavut.

The board's mission statement reads: To safeguard the caribou of the Beverly and Qamanirjuaq herds for traditional users who wish to maintain a lifestyle that includes the use of caribou, as well as for all Canadians and people of other nations.

Board activities and processes follow operating procedures developed after several years of operation. The Board itself has an advisory role. Meetings are twice a year, once in a caribou-range community and once in a central location. Decisions are generally based on consensus, but votes are taken when this is necessary. Arguably, the board's most important job relates to its recommendations for harvest allocation. A few issues, such as the commercial use of caribou, split the membership but never into government/aboriginal sides. In fact, the board is seen to work as a team with common objectives. Aboriginal people and government representatives alike have become full partners in managing the caribou resource.

Accomplishments

Of all the strides made throughout the board's history, none is more important than the improved level of trust and respect among different aboriginal and government groups that these meetings have fostered. Before, relations were uneasy as different cultures and knowledge systems collided. But both sides have made tremendous efforts to find common ground, in order to conserve caribou for the use of future generations.

Aboriginal people demonstrated remarkable courage in abandoning past positions such as giving permission to satellite-collar caribou (previously they had maintained that permission would never be given because such tampering of the animal was seen as disrespectful, interfering with the natural cycle of the caribou's life).

Government board members have learned to be patient. They have learned to understand that the aboriginal communities tend to avoid snap decisions and generally prefer decision by consensus. They prefer prolonged discussion that includes the elders in each community. As it happens, most of the BQCMB's decisions come about by consensus, with all board members agreeing on a recommendation they feel comfortable with. Often board members from caribou-range communities return home and talk with local elders and others first before bringing those viewpoints back to the next BQCMB meeting for the final decision.

Much of the board's energy is devoted to informing others about caribou conservation. Board members attend conferences across Canada to explain the BQCMB's work. Hunters who are board members report back to their communities, and board members employed by government departments advise their colleagues about BQCMB positions on different matters. In recent years, a board representative testified before a parliamentary standing committee on co-management.

Conserving the Beverly and Qamanirjuaq herds is the board's reason for being, but educating the public about caribou conservation is a major part of what the BQCMB does. Since 1982, this work has manifested itself in different, tangible ways. Following public consultation, the board published a management plan in 1987 for its first 10-year mandate, and a revised version in 1996 addressing its second 10-year mandate. The plan provides a detailed framework for management decisions, and calls upon effective teamwork and patience from all groups involved in order to protect the precious caribou resource.

Early in its mandate, the board recognized that competing land uses and activities on the ranges were based on economics, mainly mining; oil and gas; transportation corridors; and resource funding shifting to 'southern' priorities such as forestry, agriculture and other wildlife species. The board captured the attention of politicians and senior bureaucrats alike, when it published its economist-confirmed value of the subsistence harvest: over CA$11,000,000, annually. In February 2004, a technical subcommittee of the board revised this estimate upwards to over CA$17,000,000 annually, based on harvest figures, meat replacement costs, non-resident outfitting[2] (CA$3,400 per animal) and caribou by-products such as clothing and handicrafts.

In 1998, the BQCMB updated its methodology for determining allowable harvest of caribou. Harvest levels for the different category of users are fixed according to the size of the population and its ability to sustain the use. The categories are, in descending order of priority:

1   traditional users – domestic use
2   residential users – domestic use
3   traditional users – inter-settlement trade
4   local use for commercial purposes
5   outfitting
    i   traditional users
    ii   non-residents.

In 1997, the board recognized that it needed to confirm its own internal policy and procedures for recommending allocation quotas to the ministers of wildlife, who regularly seek advice on allocation priorities and numbers. The following are key statements developed by the board:

- 'The allocation of the use of caribou shall be based on the ability of the caribou population to sustain the use. The use of caribou that can be sustained shall be allocated on the basis of priority of use.'
- 'At or below the crisis herd level (150,000) as defined in the Management Plan, only Traditional Users shall be considered for allocation of caribou'
- 'Prior to allocating caribou in any priority category, demand shall be met in all priority categories higher'.
- 'When sustainable harvest decreases, or demand in higher priority categor(ies) increases, the Board shall recommend the use of caribou be removed, beginning at the lowest priority category, and proceed in sequence of increasing priority level'.
- 'Local preferences may reprioritize use categories within the non-resident uses, for example, Local Use for Commercial Purposes ahead of Traditional/ Resident Uses (Guiding)'.
- 'In the absence of domestic harvest data…the following caribou herd population thresholds for caribou use are as follows:

Use	Threshold (5%)	Harvest	Minimum % for domestic hunt
Commercial (export)	300,000	15,000	75
Commercial (non-export)	250,000	12,500	80
Resident (licensed)	200,000	10,000	90
Domestic	< 150,000	< 7500	100

The latest commercial tag allocation of 1,585 Qamanirjuaq and 425 Beverly Caribou (November 2003) was broken down by community as follows:

Baker Lake	125 Qamanirjuaq	Non-resident hunting
	35 Beverly	Meat sales
	15 Beverly	Non-resident hunting
Chesterfield Inlet	150 Qamanirjuaq	Meat sales
Rankin Inlet	145 Qamanirjuaq	Non-resident hunting
	15 Qamanirjuaq	Meat sales
Whale Cove	145 Qamanirjuaq	Meat sales
	5 Qamanirjuaq	Non-resident hunting
Arviat	200 Qamanirjuaq	Meat sales
	400 Qamanirjuaq	Non-resident hunting
Ft. Smith	200 Beverly	Meat sales
	75 Beverly	Sport hunting (unused)
Lutsel'ke	100 Beverly	Sport hunting
Manitoba	400 Qamanirjuaq	Sport hunting

The Rankin Inlet Hunters and Trappers Organization requested board support for internal reallocations of the commercial quota. The board agreed in principle, provided the overall allocation was not exceeded. However, it is a concern that jurisdictions need to track actual use of allocations, so reallocations can be made for unused tags.

The board also recommends that trophy fees be dedicated budget allocations from the jurisdictions for conservation development and management (as is done with waterfowl stamps) rather than buried in general revenue. In Nunavut, for example, the CA$150 trophy fee for 200 animals would yield CA$30,000 in funding.

As part of its ongoing work, the board has recently undertaken a number of outreach initiatives. Since 2000, the BQCMB's Caribou Monitoring Project has interviewed 16 elders and 80 hunters from Arviat and Baker Lake for their observations about caribou out on the land. The board also worked with area hunters to hammer out a 'hunter's code of ethics' – something of a moral manual for the hunting profession that crosses provincial and territorial borders. A 'Hunting Wisely' poster was also distributed to caribou-range communities in 1998. As part of more general outreach activities, the BQCMB has been distributing a small newsletter, Caribou News in Brief, since 1997. The newsletter has been well received and is published twice annually, complementing updated information given through the board's web site.

Challenges into the next decade

The BQCMB is preparing for the future, mindful that pressure must be maintained for budgets. It seeks to give attention to northern and aboriginal issues and opportunities. Board profile and liaison, and relevance to communities, especially youth, are all factors to be taken into account, with consideration to the need for continuity with the membership. The board received an early wake-up call in 2004, when one jurisdiction attempted to cancel the 'grant', another refused to send the full complement of members and several jurisdictions collectively pointed to difficulties in budgeting for caribou surveys. The board continues to press governments, land claim organizations and the private sector agencies about the impossibility of managing the herds without population and harvest data.

For their part, the community representatives continue to wrestle with promoting board programmes, and ensuring that there is support for them – most notably caribou collaring. Their concerns include such issues as:

- incursions into traditional territories from new hunters (treaty and non-treaty);
- improved access and industrial development;
- habitat loss and degradation;
- loss of interest in traditional pursuits like hunting, trapping and generally living off the land.

Government members are constantly reminded of the board's arm's-length relationship with governments, so members can 'free-wheel' rather than merely just propose programmes and policies that they believe will be acceptable to the agencies.

Summary and conclusions

The board can boast of some significant achievements that demonstrate its effectiveness as a unit. From its initiation in 1982, caribou management plans were developed cooperatively in 10-year increments. It has effectively communicated with its key constituents through meetings, community consultations, newsletters and videos. The next generation of caribou hunters and managers is being educated through a curriculum-based schools programme and poster/prose contests, as well as educational support from a board-generated scholarship fund. The board has enjoyed a good reputation amongst communities in the north and governments in the south through community skills competitions, regional, national and international presentations and parliamentary testimony, and timely and contemporary recommendations to governments. And it has solidified its standing as the pre-eminent forum for research and policy recommendations on caribou management with the sponsorship of scientific studies and special reports, economic evaluation of the caribou resource, allocation policy and procedures, and protection of the caribou range.

The board faces several challenges into the next decade. Climate change and escalating industrial development on the range for mineral exploration and exploitation pose a possible threat to caribou habitat. New roads will mean improved access to caribou and their habitats, which in turn may create pressure for increased harvests. The demands of caribou data collection, including population and harvest surveys, will continue, as will the need to further develop and maintain community support for board initiatives. Finally, there will be ongoing administrative challenges, including maintaining budgets and planning for board member continuity and succession.

The BQCMB's role is a dual one, both to safeguard the herds primarily for aboriginal people, whose lifestyle includes the use of caribou, and also to provide benefits for Canadians and people of other nations. In this respect, the board continues to be a model for effective natural resources co-management on the world stage. The five governments that fund the board have largely followed its recommendations, using the board as a forum and single window for consultation, community-based programming, coordinated management and advice. Of all the strides made, none is more important than the improved level of trust, respect and teamwork, resulting from effective and meaningful dialogue, often at the community level. Blending of scientific and traditional knowledge, experience and insight has yielded unique, effective and credible input to caribou management.

Notes

1 Looking back, it is recognized that the population estimates that gave rise to the perceived crisis might have been based on inadequate or inaccurate surveys. While the initiation of the board was a key step in improving the management of the herds, had traditional ecological knowledge been incorporated with the science into assessments of herd health policy would have been driven less by a crisis narrative and more on the basis of evidence.

2 This is the estimated economic contribution made from allowing outside hunters to take a certain number of animals per year, i.e. would include money spent on paying local guides, trophy fees, accommodation, etc.

PART 4

Regional Perspectives

Glyn Davies and Ruth Whitten

In this final part of the book, we look beyond the West and Central African region that has been the focus of our debate, to consider Asia. Here the wildlife trade has rather different underlying causes, especially in relation to the consumption of wildlife for medicinal purposes, and Elizabeth Bennett has given an important framework for considering the differences between the two regions in the chapter that follows (Chapter 15).

Here we consider worldwide variation in the availability of fish and domestic meat, which would be expected to have some influence on patterns of bushmeat trade and consumption, as indicated from one recent case study in Ghana (Brashares *et al.*, 2004). The differences between regions do offer some interesting distinctions, although the impact of fish and domestic meat availability and pricing on domestic meat production is likely to be complex and difficult to predict.

In order to look at some coarse-grained geographical patterns, data were taken from the Food and Agriculture Organization website (faostat.fao.org) to consider general trends in fish and domestic meat supplies over the last few decades. Countries that reflected the focus of the conference were selected: six from wet-zone Africa (Cameroon, DR of Congo, Gabon, Ghana, Nigeria, Sierra Leone) and three from dry-zone Africa (Tanzania, Zambia and Zimbabwe). These were compared with two countries from South-East Asia (Indonesia and Vietnam) and two from Latin America (Bolivia and Brazil).

FAO defines domestic meat supply as production and stocks, plus imports, minus exports. The published figures are provided through government departments, without independent verification by FAO in most cases, so there is a margin of error that needs to be taken into account. Despite these limitations, this overview of 13 countries did give some insight into the potential impact of domestic meat and fish availability.

Two interesting patterns arose from the global figures: undernourishment per capita was lowest, and declining, in South-East Asia and Latin America, whereas the situation was twice as bad in West Africa and three times worse in dry-zone Africa (Figure 1). The pressure to seek food from all sources, domestic or otherwise, is therefore greater in Africa than anywhere else. This distinction is likely to be exacerbated by the converse situation for domestic meat supply per capita, which clearly shows an abundance of beef, poultry

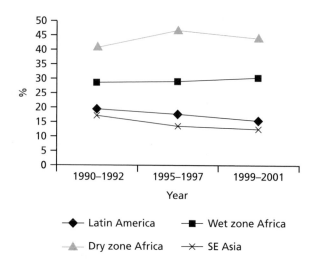

Figure 1 **Proportion of undernourishment in the four regions.**

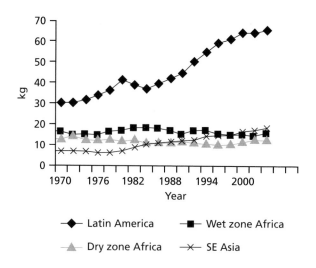

Figure 2 **Domestic meat supply per capita for the four regions.**

and pig meat in Latin America (Figure 2). Meat production in the other three regions was three times lower, although the two South-East Asian countries showed an increase in domestic meat production in recent years, with a very rapid rise in pork production in Vietnam.

Focusing on Africa, an assessment of fish availability, as an alternative source of protein, was examined from three main sources: marine capture fisheries, freshwater capture fisheries and aquaculture. Marine fish production for wet-zone Africa (West Africa) accounted for 3–4 times more food weight per capita than any other item, and is clearly a very important food source. Freshwater fish is another important source, contributing to food weight per capita as much as poultry, and more than either beef or pork. In dry-zone Africa, marine fish production crashed in the 1980s and freshwater production is much more important than poultry or pork (Figure 3). Like beef production, production of marine fish is also declining.

Latin America has relatively low levels of production of either marine or freshwater fish, but they are still above African levels, except for freshwater fish in dry-zone Africa. South Asia shows a steady increase in marine production, which outweighs the other food commodities in importance and, although freshwater fish lags far behind, it is also increasing, supported by strong output from aquaculture.

Linking these general trends with demand for bushmeat is difficult, without researching details of cause and effect on patterns of food choice and trade. One case study in Ghana has indicated that declines in wildlife populations are positively correlated with shortages of marine fish in the coastal region (Brashares *et al.*, 2004), and if this proves to be a general rule then demand for bushmeat is likely to increase if fish and domestic meat supply declines, as is currently the situation in Africa, especially those dry-zone countries considered.

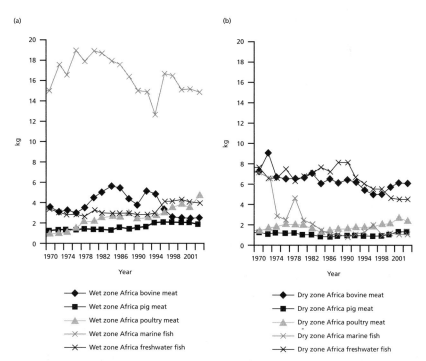

Figure 3　**Domestic supply per capita for (a) wet- and (b) dry-zone Africa.**

Much more research is needed to understand the impact of commodity availability and pricing on the bushmeat trade, especially when making inter-continental comparisons. Furthermore, different cultures, economic conditions, species and livelihood strategies occur, as we will see in the next chapter.

15

Hunting, Wildlife Trade and Wildlife Consumption Patterns in Asia

Elizabeth L. Bennett

Introduction

On average, Asia has more people and less forest than other parts of the humid tropics. In addition, parts of Asia have a long tradition of consuming wildlife for its medicinal value. Hunting and commercial wildlife trade have devastated wildlife populations across much of the region and, as a result, the livelihoods of traditional forest-dwelling peoples are threatened by loss of resources. Asia is a continent of contradictions and paradoxes, because it also contains areas with high densities of poor people whose traditions lead them to revere wildlife. In such areas, spectacular wildlife populations thrive in reserves surrounded by large numbers of impoverished people, suggesting that wildlife conservation and support for human livelihoods are not always mutually incompatible.

This chapter focuses specifically on consumption of wild meat from closed tropical forests, as the productivity of such forests for wild meat is much lower than that of other tropical ecosystems (Robinson and Bennett, 2004), so the potential of forest wildlife populations to support human livelihoods is limited, and the balance between wildlife conservation and livelihood support is concomitantly difficult to attain. The chapter reviews the different patterns of wildlife consumption across Asia, the scale and dynamics of such consumption, and the impacts of unsustainable exploitation both on wildlife populations, and also on rural peoples. The chapter concludes with a short overview of the implications for wildlife management and local livelihoods in Asia, and a comparison of the situation in Asia with that of Africa.

Consumption of wildlife in Asia

Consumption of wildlife in Asia today occurs on a vast scale. Asia has more people per unit area of remaining forest than other parts of the humid tropics: approximately 1,700 and 239 people per km^2 of forest in South and South-East Asia respectively, compared with 32 and 39 in Central Africa and tropical Latin America respectively (data from World Bank, 2001). This is in the context in which closed tropical forests can support only an estimate of one person per km^2 if they depend exclusively on wild meat for their protein (Robinson and

Bennett, 2000). In Asia, therefore, the potential supply of wildlife per capita is less than in other parts of the humid tropics.

In addition, average demand is extremely high, as wildlife is consumed both by people living in South-East Asia themselves and also by those living in other parts of Asia. Even though it is a spectrum, consumers across Asia can broadly be placed into one of three categories: rural forest dwellers, urban dwellers in habitat countries and urban dwellers in non-habitat countries.

Rural forest dwellers

Many people live in and near to Asia's forests, and forest resources play a significant role in their livelihoods. These include remote peoples who depend on wildlife for subsistence, and also cultures in transition, who are increasingly part of a developing economy, but for whom hunting is still an important part of their livelihoods.

Subsistence hunters across Asia are a significant group of consumers, even though their numbers are relatively few and declining as Asia develops. Such groups include the remaining semi-nomadic and nomadic Penan in Borneo, the Agta in the Philippines, and various hill tribes scattered in the remaining remoter forest areas of Myanmar and Indochina. Such people depend on wildlife for food. For example, 30.5% of meals of settled Penan at two communities in the Baram, Sarawak, Malaysia, contain wild meat (Chin, 2002). Also in Sarawak's highlands, prior to the arrival of logging, 67% of all meals of the Kelabits contained wild meat, and it was their main source of protein (Bennett et al., 2000). In central Sulawesi, Indonesia, the Wana eat an average of 38 g of wild meat per person per day (Alvard, 2000). And in 17 rural villages in north Sulawesi, 36% of all meat eaten is from the forest (Lee, 2000).

Data are scarce, but those from Sarawak show that many more people depend on wildlife as a source of food than as a source of income. Although most peoples throughout the interior have a high proportion of wildlife in their diet, only some of the most remote Penan groups depend on it for income. An estimated 25% of Sarawak's population, or 500,000 people, use wild meat as a significant source of protein (Bennett et al., 2000), but the number who depend on it for a significant source of income is a maximum of half of the Penan, or about 5,000 people – only 1% of the total number of people using wildlife for their protein. Virtually all wildlife sold by the Penan is sold to neighbouring communities within walking distance of the hunt; almost none makes its way into a long-distance trade. The Penan most dependent on selling wildlife tend to be those people just entering a cash economy for the first time as the development frontier hits them, often in the form of a logging road, so they start to need cash for the first time, and have few commodities apart from wildlife which they can sell. Until a recent legal trade ban, many others in Sarawak also sold wildlife, but merely as one of a number of commodities sold and incomes earned, and wildlife generally did not provide a significant proportion of a family's income.

The picture in southern Lao PDR is similar, with patterns of wildlife sales changing as the development frontier spreads. Until recently, there was little wildlife trade, with most villages being isolated and largely outside mainstream market economies. By 1990, however, the Lao Theung people were deriving 30–60% of their income from selling wildlife and other forest products (Srikosamatara *et al.*, 1992).

Across South-East Asia, although hunting is still a significant activity for many cultures in transition, wildlife rarely provides significant income once people become fully integrated into market economies. This is partly because market integration creates alternative jobs and livelihoods that are more productive than selling wildlife. Moreover, the spread of roads and commercialization of wildlife usually result in the resource being depleted to the point that it can no longer provide a viable source of income.

The economic importance of hunted wildlife to rural forest dwellers away from ready access to markets can be high, however. In Sarawak, in the mid-1990s, each rural family away from the coast ate about 313 kg of wild meat per year. The cost of replacing that with domestic pork would have been about US$1,000 per family per year, or US$75 million per year for all rural consumers of wild meat (WCS and Sarawak Forest Department, 1996). If these people could not hunt, many would farm meat rather than buy it. But the figures show the subsidy to rural economies provided by wild meat that would be lost if wildlife were no longer there.

Urban dwellers in habitat countries

Urban consumption of wildlife in countries where the animals were hunted has increased in past 25 years, as wildlife supplies have increased with the spread of logging and other roads, and as urban wealth has increased. In north Sulawesi, for example, between 1948 and 1970, only one wild meat trader was operating; two more became active from 1970 to 1984; by 1993, 12 full-time dealers were operating, and that number increased to 30 in 1996 (Clayton and Milner-Gulland, 2000). All were Minahasans, who are not traditional forest dwellers and hunters. A similar pattern was seen in Kuching, Sarawak. In 1984, the only place that consistently sold wild meat was one restaurant out of the city, and usually some could be bought in the weekend street market. By 1996, when logging roads had spread throughout the interior, almost every Chinese restaurant in the city was offering wild meat for sale every night of the week.

The scale of the domestic wild meat trade in South-East Asia is considerable. For example, in Sarawak, the wild meat trade in 1996 was conservatively estimated to be more than 1,000 tonnes per year, and mainly comprised the large ungulates: bearded pigs (*Sus barbatus*), sambar deer (*Cervus unicolor*) and barking deer (*Muntiacus* spp.) (WCS and Sarawak Forest Department, 1996). In a single market in north Sulawesi, an estimated 3,848 wild pigs were sold every year from 1993 to 1995, of which a third to a half were the endangered and legally protected babirusa (*Babyrousa babyrussa*). Annual market sales also included 50–200 macaques, 50,000–75,000 forest rats, up to 15,000 bats and occasional sales of rare cuscus and tarsiers (Clayton and Milner-Gulland, 2000). In a single food market at That Luang in southern Laos in 1992, sales involved 8,000–10,000 mammals, 6,000–7,000 birds, and 3,000–4,000 reptiles (Srikosamatara *et al.*, 1992).

Town markets are supplied from different areas over time, as any one area experiences boom and bust exploitation. For example, the wildlife markets of Minahasa District, north Sulawesi, are supplied with wildlife coming from further and further west then south, as wildlife populations in Sulawesi's northern arm are depleted (R. Lee, personal communication). In Sarawak, wildlife populations adjacent to roads are depleted rapidly as a road comes into an area. Hence, until the ban on wildlife trade in 1998, any one area of forest supplied wildlife to markets for only a short time when a road was relatively new. Fresh sources of supply followed the ends of roads as they spread.

Across Asia, urban dwellers do not depend on wildlife for nutrition, domestic meat alternatives having been developed long ago. Many Asian countries, especially those in South-East Asia, have long coastlines for their land area. Historically, centres of population concentrated along the coast. Even today, 80% of the capital cities of South-East Asia are within 100 km of the coast, as are 67% of those of South Asia. Thus, throughout Asia, marine fisheries provide a major source of protein, especially for urban populations. Traditionally, wild meat is not a significant source of protein for urban peoples. This is partly because of religious taboos amongst Hindus and Moslems, but also because urban tropical Asia has tended to look to the sea for its traditional protein. In recent years, seafood has been greatly supplemented by intensive commercial farming of livestock, especially pigs and chickens, as well as by aquaculture.

In urban areas across South-East Asia, the wildlife sold in restaurants and markets is a luxury item, often more difficult and expensive to obtain than seafood and domestic protein, and generally eaten by the wealthier members of society. It is not a basic protein source. In poorer countries such as Lao PDR, the only landlocked tropical forest Asian country, wild meat might be a supplement to urban diets, but generally this is not the case, and even here it is not a major source of protein for town people. Indeed, in parts of Indochina such as Lao PDR and Cambodia, so little wildlife remains within large distances of towns that even small birds cannot be found. Thus, although urban dwellers across South-East Asia often like to eat wildlife, and do so when it is available, its contribution to urban diets is negligible.

In South Asia, many people are vegetarians for religious reasons, so they obtain their protein from vegetables such as pulses.

Urban dwellers in non-habitat countries

Much of South and especially South-East Asia's wildlife enters international trade chains and is consumed in urban areas hundreds or thousands of miles from where it was hunted. In general, wildlife flows from the south, where it is hunted, to areas in Asia further north, where it is consumed, core consuming nations in Asia including China and Vietnam. The boundary between nutritional and medicinal consumption is blurred, with wildlife being eaten because it is good both for nutrition and for overall health. Across much of East Asia, the traditional culture to consume wildlife has now become a fashionable luxury, with the scale and impacts of the trade increasing as formerly closed countries become more open to trade, and as their populations become wealthier.

The current scale of the wildlife trade for consumption in East Asia is vast. At least 5,000 tonnes of live wild-caught turtles (TRAFFIC South-East Asia data) and 4 million tonnes of frogs' legs (Wildlife Conservation Society Indonesia data) are exported from Indonesia to China annually. The value of wildlife exports from one province in Lao PDR to Vietnam is about US$3.6 million annually, including pangolins, cats, bears and primates (Noorden and Claridge, 2001). About 1,500 restaurants in Ho Chi Minh City serve wildlife meat, and in four restaurants alone in Da Nang city over a tonne of wildlife meat is sold each week (Roberton *et al.*, 2004). Also in Vietnam, one wildlife trader in Da Nang city trades up to 6 tonnes of wildlife each week, selling to Hanoi, Ho Chi Minh City, Da Nang and China (Roberton *et al.*, 2004). Food markets in two provinces in southern China sell at least 39 species of mammals, 453 of birds, and 154 of reptiles; more than 90,000 snakes and 24,000 turtles can be seen on single market visits (Kadoorie Farm and Botanic Garden, 2004).

Effects of large-scale urban trade

On wildlife populations

The major commercial wildlife trade is causing wildlife population declines, local extinctions and the threat of global extinctions across Asia. For example, in the last 40 years, 12 species of large animals have become extinct or virtually extinct in Vietnam due mainly to hunting and wildlife trade (Bennett and Rao, 2002). In Tangkoko Duasudara Nature Reserve, north Sulawesi, Indonesia, between 1978 and 1993 the number of crested black macaques (*Macaca nigra*) declined by 75%, anoa (*Bubalus depressicornis*) and maleo birds (*Macrocephalon maleo*) by 90%, and bear cuscus (*Ailurops ursinus*) by 95% (Kinnaird and O'Brien, 1996). Across Sarawak, densities of bearded pigs, sambar deer, primates and hornbills are all significantly correlated with degree of access; the more accessible an area is, the less wildlife that remains. For example, Kubah National Park comprises intact forest but is highly accessible from the capital city, Kuching. Here, all primates and hornbills have been hunted out, and large ungulates are extremely scarce (Bennett *et al.*, 2000). In parts of Indochina, so few large animals remain that the only animals sold in markets are small birds, bats and frogs.

Trade patterns change rapidly with time, as wild stocks are depleted. In 2000, 25 tonnes of turtles were exported from Sumatra to China every week (Shepherd, 1999). By 2003, the number of turtles in the wild had declined so rapidly that exports had declined to 7 tonnes per week (WCS and TRAFFIC, 2007).

On rural forest peoples

In much of Asia the economic, social and cultural conditions are such that, once wildlife densities in the forest decline to levels where they no longer support significant hunting, people tend automatically to switch to eating other sources of protein. As their forests are opened up by roads, the wildlife disappears, usually because commercial hunting ensues. However, the road also allows people to make the dietary switch from wild to domestic forms of protein, either because they can buy domestic protein directly or because they can buy the seed stocks and feed supplies to be able to rear their own livestock. In the Bidayuh village of Mantung Merau, Sarawak, for example, within one generation the community became easily accessible by road, much of the forest was replaced by oil palm and populations of all large-mammal species declined to extremely low levels. Wild meat is now eaten in only 6.3% of meals. Proximity to outside markets, however, and many of the inhabitants now having cash-earning jobs, means that 53.3% of all meals contain non-wild protein, and people are clearly well nourished (E. L. Bennett and A. J. Nyaoi, unpublished data).

Exceptions to this are the more remote forest peoples. When roads come to their area, wildlife is hunted by outsiders, wild meat and other animal products are traded out of the areas, and wildlife populations decline. Wild meat intake usually declines. Local forest people often turn to wild fish, but the productivity of inland tropical rivers is also low; wild fish stocks are also depleted as outsiders come to the area, and sustainability of such fishing in the long term is also unlikely. Attempts at livestock rearing in such communities are not always successful, which means that people remain tied to declining wild meat supplies. Correlates of a seeming inability of remote forest peoples to adapt to declining wild meat stocks include (adapted from Langub, 1996):

1   Remoteness from markets.
2   A lack of cash coming into the community. This might be because of a lack of a local cash economy, lack of cash crops, the fact that no members of the community work in lucrative jobs elsewhere and send money home and a lack of government or other subsidies.
3   Lack of neighbouring communities of the same or other ethnic groups with whom they have a good relationship and from whom they can easily learn farming skills.

4   Cultural difficulties. For example, traditionally the Penan in Sarawak did not eat domestic animals because they considered them to be unclean.

The more marginalized peoples who meet these conditions are likely to be the ones to suffer from malnutrition as their forests are opened up to trade and wildlife declines (Robinson and Bennett, 2002). Loss of forest wildlife to urban markets presents a food security problem, therefore, for the most marginalized forest peoples who cannot easily adapt to alternative sources of protein. For example, the hunting success of a Penan community in Sarawak living close to a road was only 15% of that of a nearby community 6.5 km from the road. Wild meat was in 30.5% of the meals of the community far from the road, but only 7.9% of meals in that close to the road (Chin, 2002). In the Philippines, between 1975 and 1985, as the land of the Agta was opened up by roads and hunting pressure increased, the proportion of successful hunts declined from 63% to 16%, and the number of kills per hunt declined by 86%. The Agta went from being hunters of abundant wildlife in primary forests to being struggling foragers with little wildlife to hunt (Griffin and Griffin, 2000).

Loss of wildlife, therefore, causes nutritional hardship to the most remote, marginalized forest peoples. Moreover, the high human population densities in relation to remaining forests and wildlife supplies across much of Asia mean that wildlife resources can never be a sustainable source of income for anything beyond a tiny minority of the overall population.

Implications for management

Some hunting in Asia is for subsistence; for remote forest peoples, this is still essential. In addition, some local sales are important sources of income for a small number of such people. In general, however, most hunting in Asia today is to supply a massive, long-distance commercial market. This is driven not by the needs of poor people for cash, but by the demand of generally wealthy urban dwellers. Local hunters do benefit in the short term from selling wildlife, but the duration of significant financial benefits coming to any one rural hunter is almost inevitably extremely short as the resource is depleted so rapidly. The source of supply moves rapidly across the landscape as wildlife in any one area is depleted, with mobile middle-men obtaining wildlife from ever-changing sources to supply the major urban markets. The trade is devastating wildlife populations, and threatening the livelihoods of some of Asia's most marginalized, poor people.

For management, therefore, we should distinguish between two types of trade:

1   Local sales within rural areas, where remote hunters sell wildlife to their rural neighbours, generally within a day's walk from site of the

hunt to point of sale. Such trade is generally low in volume, and can be an important source of income for remote forest hunting communities. Traditionally, this would have been a barter trade, but now cash is usually involved.

2   Long-distance, high-volume, generally capitalized trade involving middlemen, with wildlife sold tens of kilometres or more from the point of hunt. Such trade is defined here as commercial wildlife trade, and it is this which threatens wildlife populations, and hence the livelihoods of remote forest peoples.

A management strategy throughout much of South-East Asia, therefore, aims to control or ban the long-distance commercial wildlife trade. This causes no nutritional hardship in urban areas because people there do not depend on wild meat for protein. It greatly alleviates the pressure on wildlife in the remaining forests, and also helps to conserve wild meat supplies of the forest people who truly depend on it.

Such a trade ban was put into effect in Sarawak. The 1998 'Wild Life Protection Ordinance' bans the sales of all wildlife and wildlife products taken from the wild. It has been put into effect by intensive programmes of education and enforcement, and has received strong support from rural community leaders who see it as conserving the resources on which their rural constituents depend (Bennett and Tisen, 2001). To recognize the needs of the rural Penan to sell wildlife, the trade ban is specifically implemented in urban centres, along main transport routes and in logging camps. Thus, local sales of wildlife within the interior continue, and there are no plans to prevent them.

Other examples of such wildlife management approaches are currently being implemented in north Sulawesi, where government road checks are greatly reducing the wild meat being transported to the markets of the north-east, with the aim of conserving wildlife resources in central Sulawesi (R. Lee, personal communication), and in Lao PDR, where the government is starting to strictly enforce its wildlife laws in Vientiane (A. Johnson, personal communication). Both programmes also involve working with local hunting communities, to monitor the effects of the controls.

In South Asia, the strong cultural tradition of respecting wildlife means that the philosophy of banning commercial wildlife sales has long been the case, and is generally strictly enforced. In some areas, especially north-east India, hunting is important in culture and tradition, but it is not a food security or income dependence issue.

In addition to controlling trade, management interventions needed for the most marginalized forest peoples include supporting systems that allow them to continue hunting for their livelihoods while preventing outsiders from hunting in their areas, and working with them to ensure that hunting of all species is sustainable. If sustainability is not possible, for example in areas of high human population densities, alternative sources of protein and income should be developed.

Management interventions are also needed for the cultures in transition. They tend to be at a stage where they still depend on the wildlife resource to at least some extent, but are part of a developing economy, so are often destroying the resource. If the wildlife resources of the area are important either for conservation or for supporting livelihoods, then such areas should be the focus of conservation and development efforts.

Conclusions

The picture in the humid tropics of Asia is, in many ways, different to that in tropical forest Africa, but the difference is primarily one of scale. Higher human population densities per unit area of remaining forest, and the strong traditional links between wildlife consumption and medicine, mean that pressures on Asia's wildlife are especially high. Some core commonalities exist across forested Asia and Africa: the existence of different patterns of consumption and trade within a country, often involving rural necessity and urban luxury; forest-based peoples whose livelihoods depend on conserving a wildlife resource which is threatened by long-distance trade; and overexploitation threatening the survival of wildlife species, especially those dependent on closed forests.

Given these complexities, across both continents, planning must be preceded by determining whose livelihoods are the focus of management, which species are hunted, and in which habitats. Diverse management interventions are required for the different ecosystems and types of consumer, as the approach is very different for unproductive, closed forests than more productive open systems, and is also different for truly forest-dependent people, cultures in transition and urban luxury consumers. In Asia, major efforts are needed to control the large-scale urban trade, but in both continents much effort also needs to focus on balancing conservation and development planning for forest peoples, and cultures in transition. Also in both continents, efforts are needed to ensure that wildlife populations are not threatened by unsustainable hunting, both of endangered species and of species that provide critical livelihood support. Crucially, in all programmes, irrespective of continent, the aims of interventions must be clear and unambiguous, and wildlife populations and rural livelihoods must be monitored, to ensure that management programmes are achieving their aims. Only with such clear thinking, planning and management will wildlife populations across the humid tropics be conserved, and rural livelihoods sustained.

Acknowledgements

I would like to thank Glyn Davies for asking me to present this paper at the conference, and John Robinson for his insights and comments on the paper.

References

Adams, W.M. (2004) *Against Extinction: the Story of Conservation*. London: Earthscan.

Adams, W.M. and Mulligan, M. (2003) *Decolonising Nature: Strategies for Conservation in a Postcolonial Era*. London: Earthscan.

Addo, F.A., Asibey, E.O.A., Quist, K.B. and Dyson, M.B. (1994) The economic contribution of women and protected areas: Ghana and the bushmeat trade. In *Linking Conservation and Sustainable Development*, Munasinghe, M. and McNeely, J. (eds.). Washington, DC: World Bank, pp. 68–78.

Agrawal, A. and Gibson, C. (2001) *Communities and the Environment: Ethnicity, Gender and the State in Community Based Conservation*. Piscataway, NJ: Rutgers University Press.

Alexiades, M.N. and Shanley, P. (2004) *Productos Forestales, Medios de Subsistencia y Conservación: Estudios de Caso Sobre Sistemas de Manejo de Productos Forestales no Maderables*, Vol. 3 – *America Latina*. Bogor, Indonesia: Center for International Forestry Research.

Alvard, M. (2000) The impact of traditional subsistence hunting and trapping on prey populations: data from Wana horticulturalists of upland Central Sulawesi, Indonesia. In *Hunting for Sustainability in Tropical Forests*, Robinson, J.G. and Bennett, E.L. (eds.). New York: Columbia University Press, pp. 214–230.

Amanor, K. and Brown, D. (2003) *Making Environmental Management More Responsive to Local Needs: Decentralisation and Evidence-based Policy in Ghana*. ODI Forestry Briefings No. 3. London: ODI.

Anadu, P.A., Elamah, P.O. and Oates, J.F. (1988) The bushmeat trade in southwestern Nigeria: a case study. *Human Ecology* **16**: 199–208.

Angelsen, A. and Wunder, S. (2003) *Exploring the Forest Poverty Link: Key Concepts, Issues and Research Implications*. CIFOR Occasional Paper No. 40. Bogor, Indonesia: Center for International Forestry Research.

Anstey, S. (1991) *Wildlife Utilisation in Liberia – the Findings of a National Survey 1989–1990*. Gland, Switzerland: WWF International.

Apaza, L., Wilkie, D., Byron, E., Huanca, T., Leonard, W., Perez, E., Reyes-Garcia, V., Vadez, V. and Godoy, R. (2002) Meat prices influence the consumption of wildlife by the Tsimane Amerindians of Bolivia. *Oryx* **36**: 382–388.

Arnold J.E.M. (2002) Clarifying the links between forests and poverty reduction. *International Forestry Review* **4(3)**: 231–234.

Arnold, J.E.M. and Ruiz Perez, M. (1998) The role of non-timber forest products in conservation and development. In *Incomes from the Forest: Methods for the Development and Conservation of Forest Products for Local Communities*, Wollenberg, E. and Ingles, A. (eds). Bogor, Indonesia: Center for International Forestry Research/IUCN, pp. 17–42.

Arnold, J.E.M. and Ruiz Perez, M. (2001) Can non-timber forest products match tropical forest conservation and development objectives? *Ecological Economics* **39**: 437–447.

Ashley C. and Carney, D. (1999) *Sustainable Livelihoods: Lessons from Early Experience*. London: DFID/ODI.

Ashley, C. and Lefranchi, C. (1997) *Livelihood Strategies of Rural Households in Caprivi: Implications for Conservancies and Natural Resource Management*. Research Discussion Paper No. 20. Windhoek: Namibia: Directorate of Environmental Affairs.

Ashley, C., Start, D. and Slater, R. (2003) *Livelihood Diversity and Diversification. Understanding Livelihoods in Rural India: Diversity, Change and Exclusion*. Livelihood Guidance Policy Sheet 2.3. London: ODI.

Asibey, E.O.A. (1966) Why not bushmeat too? *The Ghana Farmer* **10**: 165–170.

Asibey, E.O.A. (1974) Wildlife as a source of protein in Africa south of the Sahara. *Biological Conservation* **6**: 32–39.

Asibey, E.O.A. (1977) Expected effects of land-use patterns on future supplies of bushmeat in Africa south of the Sahara. *Environmental Conservation* **4**: 43–49.

Asibey, E.A.O. and Child, G. (1991) Wildlife management for rural development in sub-Saharan Africa. *Unasylva* **41**: 10.

Auzel, P. and Wilkie, D.S. (2000) *Wildlife Use in Northern Congo: Hunting in a Commercial Logging Concession*. In Hunting for sustainability in tropical forests, Robinson, J. G. and Bennett, E. L. (eds.). New York: Columbia University Press, pp. 413–426.

Bahuchet, S. (1993) History of the inhabitants of the Central African Rain Forest: perspectives from comparative linguistics. In *Tropical Forests, People and Food*, Hladik, C.M., Hladik, A., Linares, O.F., Pagezy, H., Semple, A. and Hadley, M. (eds). Paris: UNESCO, pp. 37–54.

Balmford A. (1996) Extinction filters and current resilience: the significance of past selection pressures for conservation biology. *Trends in Ecology and Evolution* **11**: 193–196.

Barnett, R. (2000) *Food for Thought: the Utilization of Wild Meat in Eastern and Southern Africa*. Nairobi: TRAFFIC, East/Southern Africa.

Bennett, E.L. (2000) Timber certification: where is the voice of the biologist? *Conservation Biology* **14**: 921–923.

Bennett, E.L. and Rao, M. (2002) *Hunting and Wildlife Trade in Tropical and Subtropical Asia: Identifying Gaps and Developing Strategies*. Bangkok: WCS.

Bennett, E.L and Robinson, J.G. (2000) *Hunting of Wildlife in Tropical Forests: Implications for Biodiversity and Forest Peoples*. Environment Department papers, Biodiversity series, Impact studies. Washington, DC: World Bank.

Bennett, E.L and Robinson, J.G. (2002) Into the frying pan and onto the shelves of the apothecary: the fate of tropical forest wildlife in Asia. In *Terrestrial eco-regions of Asia: A conservation Assessment*. Washington DC: Island Press.

Bennett, E.L. and Tisen, O.B. (2001) A master plan for wildlife conservation. *Sarawak Gazette* **CXXVIII (1543)**: 4–10.

Bennett, E.L., Nyaoi, A.J. and Sompud, J. (2000) Saving Borneo's bacon: the sustainability of hunting in Sarawak and Sabah. In *Hunting for Sustainability in Tropical Forests*, Robinson, J.G. and Bennett, E.L. (eds.). New York: Columbia University Press, pp. 305–324.

Bennett, E.L., Eves, H., Robinson, J.G. and Wilkie, D. (2002) Why is eating bushmeat a biodiversity crisis? *Conservation in Practice* **3**: 28–29.

Bennett, E.L., Blencowe, E., Brandon, K., Brown, D., Burn, R.W., Cowlishaw, G.C., Davies, G., Dublin, H., Fa, J., Milner-Gulland, E.J., Robinson, J.R., Rowcliffe, J.M., Underwood, F. and Wilkie, D. (2007) Hunting for Consensus: a statement on reconciling bushmeat harvest, conservation and development policy in west and central Africa. *Conservation Biology* **21**: 884–887.

Blom, A., Almasi, A., Heitkönig, I.M.A., Kpanou, J.-B. and Prins, H.H.T. (2001) A survey of the apes in the Dzanga-Ndoki National Park, Central African Republic: a comparison between the census survey method of estimating the gorilla (*Gorilla gorilla gorilla*) and chimpanzee (*Pan troglodytes*) nest group density. *African Journal of Ecology* **39**: 98–105.

Bojö, J., and Reddy, R.C. (2002) *Poverty Reduction Strategies and Environment, a Review of 40 Interim and Full Poverty Reduction Strategy Papers*. Washington DC: World Bank.

Boot, R. and Gullison, R.E. (1995) Approaches to developing sustainable extraction systems for tropical forest products. *Ecological Applications* **5**: 896–903.

Booth, D. (2003) *Fighting poverty in Africa: are PRSPs making a difference?* London: ODI.

Bowen-Jones, E. (1998) A review of the commercial bushmeat trade with emphasis on Central/West Africa and the great apes. *African Primates*, 3, Supplement S1-S43.

Bowen-Jones, E., Brown, D. and Robinson, E. (2002) *Assessment of the Solution Orientated Research Needed to Promote a More Sustainable Bushmeat Trade in Central and West Africa*. Research report to DEFRA. Bristol: DEFRA.

Bowen-Jones, E., Brown, D. and Robinson, E.J.Z. (2003) Economic commodity or environmental crisis? An interdisciplinary approach to analyzing the bushmeat trade in central and west Africa. *Area* **35**: 390–402.

Brashares, J.S., Arcese, P. and Sam, M.K. (2001) Human demography and reserve size predict wildlife extinction in West Africa. *Proceedings of the Royal Society of London Series B Biological Sciences* **268**: 2473–2478.

Brashares, J.S., Arcese, P., Sam, M.K., Coppolillo, P.B., Sinclair, A.R.E. and Balmford, A. (2004) Bushmeat hunting, wildlife declines, and fish supply in West Africa. *Science* **306:** 1180–1183.

Brocklesby, M.A., and Hinshelwood, E. (2001) *Poverty and the Environment: What the Poor Say. An Assessment of Poverty–Environment Linkages in Participatory Poverty Assessments.* Swansea: Centre for Development Studies, University of Wales Swansea.

Brown, D. (2003a) Is the best the enemy of the good? Livelihoods perspectives on bushmeat harvesting and trade – some issues and challenges. *Proceedings of the International Conference on Rural Livelihoods, Forests and Biodiversity,* organized by CIFOR, Bonn, Germany.

Brown, D. (2003b) *Bushmeat and Poverty Alleviation: Implications for Development policy.* Wildlife Policy Briefing No. 2. London: ODI.

Brown, D. and Williams, A. (2003) The case for bushmeat as a component of Development policy: issues and challenges. International Forestry Review 5: 148–155.

Brown D., Schreckenberg, K. Shepherd, G. and Wells, A. (2002) *Forestry as an Entry Point for Governance Reform.* ODI Forestry Briefing No. 1. London: ODI.

Butynski, T.M. and Koster, S.H. (1994) Distribution and conservation status of primates in Bioko Island, Equatorial Guinea. *Biodiversity and Conservation* **3:** 893–909.

Carney, D. (1998) *Sustainable Rural Livelihoods: What Contribution Can we Make?* London: DFID/ ODI.

Carroll, R.W. (1988) Relative density, range extension and conservation potential of the lowland gorilla *Gorilla gorilla gorilla* in the Dzanga-Sangha region of southwestern Central African Republic. *Mammalia* **52:** 309–323.

Caspary, H.-U. (1999) *Wildlife Utilisation in Cote d'Ivoire and West Africa – Potentials and Constraints for Development Cooperation.* Eschborn, Germany: Deutsche Gesellschaft für Technische Zusammenarbeit (GTZ) GmbH.

Cernea, M. and Schmidt-Soltau, K. (2003) *National Parks and Poverty Risks: is Population Resettlement the Solution?* Paper presented at the World Park Congress, Durban, September 2003. [An abbreviated version is published as: The end of forced resettlements for conservation: Conservation must not impoverish people, *Policy Matters* **12:** 42–51.]

Chardonnet, P., Fritz, H., Zorzi, N. and Feron, E. (1995) Current importance of traditional hunting and major contrasts in wild meat consumption in sub-saharan Africa. In *Integrating the People and Wildlife for a Sustainable Future,* Bissonette, J.A. and Krausman, P.R. (eds). USA: The Wildlife Society, pp. 304–307.

Chin, C.L.M. (2002) *Hunting Patterns and Wildlife Densities in Primary and Production Forests in the Upper Baram, Sarawak.* Master's Thesis, University Malaysia, Sarawak, Malaysia.

CIB (2006) *Plan d'Amenagement de l'Unité Forestière d'Aménagement de Kabo (2005–2034).* CIB and MEFE.

Clark, G. (1994) *Onions are my Husband: Survival and Accumulation by West African Market Women.* Chicago: Chicago University Press.

Clayton, L. and Milner-Gulland, E.J. (2000) The trade in wildlife in north Sulawesi, Indonesia. In *Hunting for Sustainability in Tropical Forests:,* Robinson, J.G. and Bennett, E.L. (eds.). New York: Columbia University Press, pp. 473–496.

Colell, M., Maté, C. and Fa, J.E. (1994) Hunting among Moka Bubis in Bioko: dynamics of faunal exploitation at the village level. *Biodiversity and Conservation* **3:** 939–950.

Colyn, M.M., Dudu, A. and ma Mbalele, M. (1987) Data on small and medium scale game utilization in the rain forest of Zaire. In *WWF: Wildlife Management In sub-Saharan Africa: Sustainable Economic Benefits and Contribution Towards Rural Development.* Harare, Zimbabwe: World Wide Fund for Nature, pp. 109–145.

Conway, T, Moser, C, Norton, A. and Farrington, J. (2002) *Rights and Livelihoods Approaches: Exploring Policy Dimensions.* Natural Resource Perspectives No. 78. London: ODI.

Cowlishaw, G., Mendelson, S. and Rowcliffe, J.M. (2004) *The Bushmeat Commodity Chain: Patterns of Trade and Sustainability in a Mature Urban Market in West Africa.* Wildlife Policy Briefing No. 2. London: ODI.

Cowlishaw, G., Mendelson, S. and Rowcliffe, J. M. (2005a) Structure and operation of a bushmeat commodity chain in Southwestern Ghana. *Conservation Biology* **19:** 139–149.

Cowlishaw, G., Mendelson, S. and Rowcliffe, J. M. (2005b) Evidence for post-depletion sustainability in a mature bushmeat market. *Journal of Applied Ecology* **42:** 460–468.

Crookes, D.J., Ankudey, N. and Milner-Gulland, E.J. (2006) The usefulness of a long-term bushmeat market dataset as an indicator of system dynamics. *Environmental Conservation* **32:** 333–339.

Damania, R., Milner-Gulland, E.J. and Crookes, D.J. (2005) A bioeconomic analysis of bushmeat hunting. *Proceedings of the Royal Society of London Series B Biological Science* **272:** 259–266.

Davies, A.G. (1987) *The Gola Forest Reserves, Sierra Leone*. Gland, Switzerland, and Cambridge, UK: IUCN.

Davies, A.G. (1990) *Crop Protection, Sierra Leone: Mammal Ecology, Crop Damage and Pest Control*. Technical Report 13. Rome: FAO.

Davies, A.G. (2002) Bushmeat and international development. *Conservation Biology* **16:** 587–589.

Davies, A.G. and Richards, P. (1991) *Rain Forest in Mende Life: Resources and Subsistence Strategies in Rural Communities around the Gola North Forest Reserve, Sierra Leone*. London: ESCOR, UK Overseas Development Administration.

Davies, A.G., Heydon, M., Leader-Williams, N., McKinnon, M. and Newing, H. (2001) The effects of logging on tropical forest ungulates. In *The Cutting Edge: Conserving Wildlife in Logged Tropical Forests*, Fimbel, R.A., Grojal, A. and Robinson, J.G. (eds). New York: Columbia University Press, pp. 93–124.

De Beer, J.H. and McDermott, M. (1989) *The Economic Value of Non-timber Forest Products in South East Asia*. Amsterdam: The Netherlands Committee for IUCN.

Delvingt, W. (1997) *La Chasse Villageoise: Synthèse Régionale des Etudes Réalisées Durant la Première Phase du Programme ECOFAC au Cameroun, au Congo, et en République Centrafricaine*. Gembloux, Belgium: ECOFAC AGRECO-CTFT, Faculté Universitaire des Sciences Agronomiques des Gembloux.

de Merode, E. and Cowlishaw, G. (2006) Species protection, the changing informal economy, and the politics of access to the bushmeat trade in the Democratic Republic of Congo. Conservation Biology 20: 1262–1271.

de Merode E., Homewood, K. and Cowlishaw, G. (2003) *Wild Resources and Livelihoods of Poor Households in the Democratic Republic of Congo*. Wildlife Policy Briefing No. 1. London: ODI.

de Merode, E., Homewood, K. and Cowlishaw, G. (2004) The value of bushmeat and other wild foods to rural households living in extreme poverty in Democratic Republic of Congo. *Biological Conservation* **118:** 573–581.

Dei, G.J.S. (1989) Hunting and gathering in a Ghanaian rain-forest community. *Ecology of Food and Nutrition* **22:** 225–243.

Dethier, M. (1995) *Etude chasse*. Yaounde, Cameroon: ECOFAC.

Dietz, T., Dolsak, N., Ostram, E. and Stern, P.C. (2000) The drama of the commons. In *The Drama of the Commons*, Ostrom, E., Dietz, T., Dolsak, N., Stern, P.C., Stonich, S. and Weber, E.U. (Eds). Washington, DC: National Academy Press, pp. 3–36.

East, T., Kümpel, N.F., Milner-Gulland, E.J. and Rowcliffe J. M. (2005) Determinants of urban bushmeat consumption in Río Muni, Equatorial Guinea. *Biological Conservation* **126:** 206–215.

Elkan, P.W., Elkan, S.W., Moukassa, A., Malonga, R., Ngangoué, M. and Smith, J.L.D. (2005) Managing threats from bushmeat hunting in a Timber Concession in the Republic of Congo. In *Emerging Threats to Tropical Forests*, Peres, C. and Laurence, W. (eds). Chicago: University of Chicago Press.

Elliott, J. (2002) *Wildlife and Poverty Study*. London: DFID Rural Livelihoods Department.

Ellis, F. and Ade Freeman, H.A. (2004) Rural livelihoods and poverty reduction strategies in four African countries. *The Journal of Development Studies* **40(4):** 1–30.

Energy Information Administration (EIA) (2003) *Equatorial Guinea*, Vol. 2003. Department of Environment. Available at http://www.eia.doe.gov/emeu/cabs/eqguinea.html (accessed 15 April 2003).

Fa, J.E. (2000) Hunted animals in Bioko Island, West Africa: sustainability and future. In Hunting for Sustainability in Tropical Forests: 168–198. Robinson, J.G. and Bennett, E. (Eds). New York, USA: Columbia University Press.

Fa, J.E. and Garcia Yuste, J.E. (2001) Commercial bushmeat hunting in the Monte Mitra forests, Equatorial Guinea: extent and impact. *Animal Biodiversity and Conservation* **24:** 31–52.

Fa, J.E. and Peres, C.A. (2001) Game vertebrate extraction in African and Neotropical forests: an intercontinental comparison. In *Conservation of Exploited Species*, Reynolds, J.D., Mace, G.M., Robinson, J.G. and Redford, K.H. (eds). Cambridge: Cambridge University Press, pp. 203–224.

Fa, J.E., Juste, J., Perez del Val, J. and Castroviejo, J. (1995) Impact of market hunting on mammal species in Equatorial Guinea. *Conservation Biology* **9**: 1107–1115.

Fa, J.E., Garcia Yuste, J.E. and Castelo, R. (2000) Bushmeat markets on Bioko Island as a measure of hunting pressure. *Conservation Biology* **14**: 1602–1613.

Fa, J.E., Peres, C.A. and Meeuwig, J. (2002a) Bushmeat exploitation in tropical forests: an intercontinental comparison. *Conservation Biology* **16**: 232–237.

Fa, J.E., Juste, J., Burn, R.W., and Broad, G. (2002b) Bushmeat consumption and preferences of two ethnic groups in Bioko Island, West Africa. *Human Ecology* **30**: 397–416.

Fa, J.E., Currie, D. and Meeuwig, J. (2003) Bushmeat and food security in the Congo Basin. *Environmental Conservation* **30**: 71–78.

Fa, J.E., Johnson, P.J., Dupain, J., Lapuente, J., Köster, P. and Macdonald, D.W. (2004) Sampling effort and dynamics of bushmeat markets. *Animal Conservation* **7**: 409–416.

Fa, J.E., Ryan, S.L., and Bell, D.J. (2005) Hunting vulnerability, ecological characteristics and harvest rates of bushmeat species in Afrotropical forests. *Biological Conservation* **121**: 167–176.

Fa, J.E., Seymour, S., Dupain, J., Amin, R., Albrechtsen, L. and Macdonald D. (2006) Getting to grips with the magnitude of exploitation: bushmeat in the Cross-Sanaga rivers region, Nigeria and Cameroon. *Biological Conservation* **129**: 497–510.

Fafchamps, M. and Gabre-Madhin, E. (2001) *Agricultural Markets in Benin and Malawi: operation and Performance of Traders*. Oxford: Centre for the Study of African Economies, University of Oxford.

Falconer, J. (1990) *The major significance of 'Minor' Forest Products: The Local Use and Value of Forests in the West African Humid Forest Zone*. Community Forestry Note No. 6. Rome: FAO.

Falconer, J. (1992) *Non-timber Forest Products in Southern Ghana*. Kent: Natural Resources Institute.

Falconer, J. (1996) Developing research frames for non-timber forest products. In *Current Issues in Non-Timber Forest Product Research*, Ruiz-Perez, M. and Arnold, J.E.M. (eds). Bogor, Indonesia: Center for International Forestry Research, pp. 143–160.

FAO (1995) *Non Wood Forest Products for Rural Income and Sustainable Forestry*. NWFPs 7. Rome: FAO.

FAO (2003) Resumen informativo sobre la pesca por paises: La Republica de Guinea Ecuatorial, Vol. 2004. Rome: FAO; available at http://www.fao.org/fi/fcp/es/GNQ/profile.html (accessed 8 October 2004).

Fay, J.M. and Agnagna, M.A. (1991) Population survey of forest elephants (*Loxodonta africana cyclotis*) in northern Congo. *African Journal of Ecology* **29(3)**: 177–187.

Fay, J.M. and Agnagna, M.A. (1992) Census of gorillas in the northern Republic of Congo. *American Journal of Primatology* **27**: 275–284.

Feer, E. (1993) The potential for sustainable hunting and rearing of game in tropical forests. In *Tropical Forests, People and Food: Biocultural Interactions and Applications to Development:*. Hladik, C.M., Linares, O.F., Pagezy, H., Semple, A., and Hadley, M. (eds). Paris: UNESCO, pp. 691–708.

Fimbel, C.C. (1994) Ecological correlates of species success in modified habitats may be disturbance- and site-specific: the primates of Tiwai Island. *Conservation Biology* **8**: 106–113.

Fitzgibbon, C. (1998) The management of subsistence harvesting: behavioural ecology of hunters and their mammalian prey. In *Behavioural Ecology and Conservation Biology*, T. Caro (ed.). Oxford: Oxford University Press, pp. 449–473.

Fontana, M., Joekes, S. and Masika, R. (1998) *Global Trade Expansion and Liberalisation: Gender Issues and Impacts*. BRIDGE Report No. 42. Sussex: Institute of Development Studies.

Freese, C.H. (1998) *Wild Species as Commodities: Managing Markets and Ecosystems for Sustainability*. Washington, DC: Island Press.

Garcia Yuste, J.E. (1995) *Inventario y Censo de las Poblaciones de Primates de Parque Nacional Monte Alen*. Republic of Equatorial Guinea: ECOFAC/AGRECO-CTFT.

Geschiere, P. (1995) *Sorcellerie et Politique: la Viande des Autres*. Paris: Karthala.

Gibson, C.G. (2001) The role of community in natural resource conservation. In *Communities and Nature*, Agrawal A. and Gibson C. (eds.). Piscataway, NJ: Rutgers University Press, p. 14.

Gibson, C. and Marks, S.A. (1995) Transforming rural hunters into conservationists? An assessment of community-based wildlife management programs in Africa. World Development **23**: 941–957.

Global Forest Watch (2002) *An Analysis of Access into Central Africa's Rainforests*. Washington, DC: World Resources Institute.

Gonzalez Kirchner, J.P. (1994) *Ecologia y Conservacion de los Primates de Guinea Ecuatorial*. Malabo, Equatorial Guinea: CEIBA Ediciones.

Government of Ghana (1998) *Wildlife Laws of Ghana*. Accra, Ghana: Wildlife Department, Ministry of Lands and Forestry.

Government of the Republic of Namibia (1995) *Wildlife Management, Utilisation and Tourism in Communal Areas*. Policy Document. Windhoek, Namibia: Ministry of Environment and Tourism.

Government of the Republic of Namibia (1996) *Nature Conservation Amendment Act. Government Gazette No. 1333*. Windhoek, Namibia.

Griffin, P.B. and Griffin, M.B. (2000) Agta hunting and the sustainability of resource use in northeastern Luzon, Philippines. In *Hunting for Sustainability in Tropical Forests*, Robinson, J.G. and Bennett, E.L. (eds.). New York: Columbia University Press, pp. 325–335.

Grimble, R. and Laidlaw, M. (2002) *Biodiversity Management and Local Livelihoods: Rio Plus 10*. ODI Natural Resource Perspectives No. 73. London: ODI.

Grubb, P., Jones, T.S., Davies, A.G., Edberg, E., Starin, E.D. and Hill, J.E. (1998) *Mammals of Ghana, Sierra Leone and The Gambia*. St Ives: The Trendrine Press.

Guyer, J.I. (1987) Feeding Yaounde, capital of Cameroon. In *Feeding African Cities: Studies in Regional Social History*, Guyer, J.I.. (ed.). Manchester: Manchester University Press, pp. 112–153.

Hart, J.A. (1978) From subsistence to market: A case study of the Mbuti net hunters. *Human Ecology* **6**: 32–53.

Hart, J.A. (2000) Impact and sustainability of indigenous hunting in the Ituri Forest, Congo-Zaire: a comparison of hunted and unhunted duiker populations. *In Hunting for Sustainability in Tropical Forests*:, Robinson. J.G. and Bennett, E.L. (eds). New York: Columbia University Press, pp. 106–153.

Hartley, D. (1993) *Forest Resource Use and Human Subsistence in Sierra Leone*. PhD thesis, University College of London.

Hasler, R. (1996) *Agriculture Foraging and Wildlife Resource Use in Africa: Culture and Political Dynamics in the Zambezi Valley*. London: Kegan Paul.

Hennemann, W.W. (1983) Relationship among body-mass, metabolic-rate and the intrinsic rate of natural increase in mammals. *Oecologia* **56**: 104–108.

Hinderink, J. and Sterkenberg, J. (1975) *Anatomy of an African Town: a socio-economic Study of Cape Coast, Ghana*. Utrecht: University of Utrecht.

Hinz, M.O. (2003) *Without Chiefs There Would be no Game: Customary Law and Nature Conservation*. Windhoek, Namibia: Out of Africa Publishers.

Hofman, T., Ellenberg. H. and Roth, H.H. (1999) *Bushmeat: a Natural Resource of the Moist Forest Regions of West Africa*. TÖB F-V/7e. 1999. Eschborn: Gesellschaft für Technische Zusammenarbeit.

Holbech, L. (1998) *Bushmeat Survey: Literature Review, Field Work and Recommendations for a Sustainable Community-based Wildlife Resource Management System*. Accra: Protected Areas Development Programme, Wildlife Department, Ministry of Lands and Forestry, Ghana.

Holzmann, R. and Jørgensen, S. (2000) *Social Risk Management: a New Conceptual Framework for Social Protection and Beyond*. Social Protection Discussion Paper Series No. 006. Washington, DC: The World Bank.

Hoyt, R. (2003) *Wildmeat Harvest and Trade in Liberia: Managing Biodiversity, Economic and Social Impacts*. Wildlife Policy Briefing No. 6. London: ODI.

Hulme, D. and Murphree, M. (2001) *African Wildlife and Livelihoods: the Promise and Performance of Community Conservation*. London: Heinemann and James Currey.

Jeffrey S. (1977) How Liberia uses wildlife. *Oryx* **14**: 169–173.

Jeffries, R. (1978) *Class, Power and Ideology in Ghana: the Railwaymen of Sekondi*. Cambridge: Cambridge University Press.

Jerozolimski, A. and Peres, C.A. (2003) Bringing home the biggest bacon: a cross-site analysis of the structure of hunter-kill profiles in neotropical forests. *Biological Conservation* **111**: 415–425.

Jones, B. (2003) The evolution of community based approach to wildlife management in Kunene, Namibia. In *African Wildlife and Livelihoods: the Promise and Performance of Community Conservation*. Hulme, D. and Murphree, M. (eds). Cape Town: David Phillips.

Jones, B.T.B. (1999) *Community Management of Natural Resources in Namibia. IIED Drylands Programme*. Issue Paper No. 90. London: International Institute for Environment and Development (IIED).

Jones, T.S. (1998) The Sierra Leone monkey drives. In *Mammals of Ghana, Sierra Leone and Gambia*, Grubb, P., Jones, T.S., Davies, A.G., Starin, E.D. and Hill, J.E. (eds). St Ives: Trendrine Press, pp. 214–219.

Juste, J., Fa, J.E., Perez del Val, J. and Castroviejo, J. (1995) Market dynamics of bushmeat species in Equatorial Guinea. *Journal of Applied Ecology* **32**: 454–467.

Kadoorie Farm and Botanic Garden (2004) *Wild Animal Trade Monitoring at Selected Markets in Guangzhou and Shenshen, South China, 2000–2003*. Hong Hong SAR: Kadoorie Farm and Botanic Garden.

Kalpers, J., Williamson E.A., Robbins, M.M., McNeilage, A., Nzamurambaho, A. Lola, N. and Mugiri, G. (2003) Gorillas in the crossfire: population dynamics of the Virunga mountain gorilla population over the past three decades. *Oryx* **37**: 326–337.

Kingdon, J. (1997) *The Kingdon Field Guide to African Mammals*. London: Academic Press.

Kinnaird, M. and O'Brien, T. (1996) Changing populations of birds and mammals in north Sulawesi. *Oryx* 30: 150–156.

Kümpel, N.F. (2006) *Incentives for Sustainable Hunting of Bushmeat in Río Muni, Equatorial Guinea*. PhD thesis, Imperial College London, University of London, and Institute of Zoology, Zoological Society of London, London.

Lahm, S.A. (1993) *Ecology and Economics of Human/Wildlife Interactions in Northeastern Gabon*. PhD thesis, New York University, New York.

Lahm, S.A. (1994) *Hunting and Wildlife in Northeastern Gabon: Why Conservation Should Extend Beyond Protected Areas*. Makokou: Institut de Recherche en Ecologie Tropicale.

Langub, J. (1996) Penan response to change and development. In *Borneo in Transition: People, Forests, Conservation, and Development:*. Padoch, C. and Peluso, N.L. (eds.). Kuala Lumpur: Oxford University Press, pp. 103–120

Leader-Williams, N. (1988) Patterns of depletion in a black rhinoceros population in the South Luangwa National Park. *African Journal of Ecology* **26**: 181–187.

Leader-Williams, N. and Milner-Gulland, E.J. (1993) Policies for the enforcement of wildlife laws: the balance between detection and penalties in the Luangwa Valley, Zambia. *Conservation Biology* **7(3)**: 611–617.

Lee, R.J. (2000) Impact of subsistence hunting in north Sulawesi, Indonesia, and conservation options. In *Hunting for Sustainability in Tropical Forests*, Robinson, J.G. and Bennett, E.L. (eds). New York: Columbia University Press, pp. 455–472.

Lewis, D. (2000) *Comparative Study of Factors Influencing ADMADE Success*. Lusaka: USAID.

Lewis, D. and Tembo, N. (2000) *A Rural Development Approach to Wildlife conservation: Putting Law Enforcement into Perspective*. An ADMADE Lessons-Learned Paper. African College for CBNRM.

Lewis, D. and Tembo, N. (2001*) Non-conventional Approaches to Wildlife Management in an African Landscape*. Paper presented at Pretoria Wildlife Utilization Conference.

Lewis, D. and Jackson, J. (2005) Safari hunting and conservation on communal land in southern Africa. In *People and Wildlife: Conflict or Coexistence?*, Woodroffe, R., Thirgood, S. and Rabinowitz, A. (eds.) Cambridge: Cambridge University Press, pp. 239–251.

Lewis, J. (2002) *Forest Hunter-Gatherers and their World: a Study of the Mbendjele Yaka Pygmies of Congo-Brazzaville and their Secular and Religious Activities and Representations*. PhD Dissertation, University of London.

LIFE (2002) *Living in a Finite Environment (LIFE) Programme: Semi annual report for the period April 1 through September 30, 2002*. Submitted to LIFE steering committee, Windhoek, Namibia.

Ling, S. and Milner-Gulland, E.J. (2006) Assessment of the sustainability of bushmeat hunting based on dynamic bioeconomic models. *Conservation Biology* Online Early doi:10.1111/j.1523–1739.2006.00414.x.

Long, A. (ed.) (2004) *Livelihoods and CBNRM in Namibia: the Findings of the WILD Project, Final Technical Report of the Wildlife Integration for Livelihood Diversification Project (WILD)*. Prepared for the Directorates of Environmental Affairs and Parks and Wildlife Management, the Ministry of Environment and Tourism, the Government of the Republic of Namibia. Windhoek, March 2004.

Lyon, F. (1999) Understanding market relations and bargaining power: farmer-trader interactions in agricultural development in Brong-Ahafo Region, Ghana. In *Natural Resource Management in Ghana and its Socio-economic Context*, Blench, R. (ed.). London: ODI, pp. 162–177.

McCullough, D.R. (1996) Spatially structured populations and harvest theory. *Journal of Wildlife Management* **60**: 1–9.

McGraw, W.S. (2005) Update on the search for Miss Waldron's red colobus monkey. *International Journal of Primatology* **26**: 605–619.

MacKenzie, C. (2002) *Bushmeat in Ghana*. Research report to DFID. London: DFID.

McSweeney, K. (2003) *Tropical Forests as Safety Nets? The Relative Importance of Forest Product Sale as Smallholder Insurance, Eastern Honduras*. Draft Paper, CIFOR-Bonn Conference.

McSweeney, K. (2005) *Forest Product Sale as Financial Insurance: Evidence from Honduran Smallholders*. Wildlife Policy Brief No. 10, London: ODI.

Mano Consultancy Services Ltd. (2001) Economic Change, Poverty and Environment Project. Zambia Phase II Summary Report Lusaka: Mano Consultancy Services Ltd.

Manu, C.K. (1987) National report on wildlife utilization in Ghana. In *Wildlife management in Sub-Saharan Africa*, de Clers, B. (ed.). Paris: Foundation Internationale pour la Sauvegarde du Gibier.

Marks, S. (1976; 2005) *Large Mammals and a Brave People: Subsistence Hunters in Zambia*, 2nd edn (with a new Preface and Afterword). New Brunswick, NJ: Transaction Publishers.

Marks, S. (2001) *Customary and Modern Wildlife Management Regimes*. Lusaka: Mano Consultancy Services Ltd.

Marks, S. (2004) Reconfiguring a political landscape: transforming slaves and chiefs in colonial Northern Rhodesia. *Journal of Colonialism and Colonial History* **5**: 3.

Marshall, E., Newton, A.C., Schreckenberg, K. (2003) Commercialising non-timber forest products: first steps in analysing the factors influencing success. *International Forestry Review* **5**: 128–137.

Marshall, E., Schreckenberg, K. and Newton, A.C. (eds.) (2006) *Commercialization of Non-Timber Forest Products: Factors influencing Success. Lessons Learned from Mexico and Bolivia and Policy Implications for Decision-makers*. Cambridge: UNEP-World Conservation Monitoring Centre.

Martin, C. (1991) *The rainforests of West Africa: Ecology, Threats, Conservation*. Basle: Birkhauser Verlag.

Martin, G.H.G. (1983) Bushmeat in Nigeria as a natural resource with environmental implications. *Environmental Conservation* **10**: 125–132.

Mayes, S. (2004) Map of Communal Area Conservancies. Windhoek, Namibia: Namibia Nature Foundation.

Melnick, M. (1995) *The Contribution of Forest Foods to the Livelihoods of the Huottuja (Piaroa) People of Southern Venezuela*. PhD Dissertation, Imperial College, University of London.

Mendelson, S., Cowlishaw, G. and Rowcliffe, J.M. (2003) Anatomy of a bushmeat commodity chain in Takoradi, Ghana. *Journal of Peasant Studies* **31**: 73–100.

Milner-Gulland, E.J. and Leader-Williams, N. (1992) A model of incentives for the illegal exploitation of black rhino and elephant: poaching pays in the Luangwa Valley, Zambia. *Journal of Applied Ecology* **29**: 337–339.

Milner-Gulland, E.J., Bennett, E.L. and the SCB 2002 Annual Meeting Wild Meat Group (2003) Wild meat – the bigger picture. *Trends in Ecology and Evolution* **18**: 351–357.

Ministério de Planificación y Desarrollo Economico (2002) *Resumen de los principales resultados del III Censo General de poblacion y viviendas en Guinea Ecuatorial*. Direccion General de Estadistica y Cuentas Nacionales, Ministério de Planificación y Desarrollo Económico, Republica de Guinea Ecuatorial, Malabo.

Morgan, D., Sanz, C., Onononga, J.R. and Strindberg, S. (2006) Ape abundance and habitat use in the Goualougo Triangle, Republic of Congo. *International Journal of Primatology* **27**: 147–179.

Moukassa, A., Poulsen, J.R., Clark, C.J., Mavah, G. and Elkan P.W. (2007) Bushmeat availability and consumption in a logging town in northern Congo. In *Bushmeat Hunting, Consumption and Trade in Central Africa*, Bennett, E.L., Deutsch, J.C. and Siex, D.S. (eds) (in press).

Muchaal, P.K. and Ngandjui, G. (1999) Impact of village hunting on wildlife populations in the western Dja Reserve, Cameroon. *Conservation Biology* **13**: 385–396.

Mulonga, S. (2003) *Wild Food: Use of Natural Resources for Food in Eastern Caprivi*. DEA Research Discussion Paper, No. 62. Namibia.

Murphy, C. and Mulonga, S. (2003) *Senior Community Field Ranger Workshop – Caprivi region*. WILD project working paper 13. Namibia.

Myers, N. (1988) Tropical forests: much more than stocks of wood. *Journal of Tropical Ecology* **4**: 209–221.

NACSO (2001) *Namibia's Community Based Natural Resource Management Programme – Enhancing Conservation Development and Democracy in Namibia's Rural Areas*. NACSO Brochure, Windhoek Namibia.

Naughton-Treves, L., Mena, J.L., Treves, A., Alvarez, N. and Radeloffs, V.C. (2003) Wildlife survival beyond park boundaries: the impact of slash-and-burn agriculture and hunting on mammals in Tambopata Peru. *Conservation Biology* **17**: 1106–1117.

Nepstad, D. and Schwartzman, S. (1992) Non-timber forest products from tropical forests. Evaluation a conservation and development strategy. *Advances in Economic Botany* **9**.

Neumann, R.P. and Hirsch, E. (2000) *Commercialisation of Non-timber Forest Products: Review and Analysis of Research*. Bogor, Indonesia: Center for International Forestry Research.

Newing, H. (2001) Bushmeat hunting and management: implications for duiker ecology and interspecific competition. *Biodiversity and Conservation* **10**: 99–118.

Ngnegueu, P. (1998) Exploitation de la faune sauvage dans la région du Dja (sud-est Cameroun). In *Rapport du séminaire/atelier sur l'exploitation durable de la faune dans le sud-est Cameroun*, Vabi, M. and Schoorl, J. (eds). WWF, Unpublished report.

Njiforti, H.L. (1996) Preferences and present demand for bushmeat in north Cameroon: Some implications for wildlife conservation. *Environmental Conservation* 23: 149–155.

Noorden, H. and Claridge, G. (2001) *Wildlife Trade in Laos: The End of the Game*. Amsterdam: Netherlands Committee for IUCN.

Norton, A. and Moser, C. (2001) *To Claim Our Rights: Livelihoods Security, Human Rights and Sustainable Development*. London: ODI.

Noss, A.J. (1995) *Duikers, Cables and Nets: a Cultural Ecology of Hunting in a Central African Forest*. PhD thesis, University of Florida, Gainsville.

Ntiamoa-Baidu, Y. (1998) *Sustainable Harvesting, Production and Use of bushmeat*. Accra, Republic of Ghana: Wildlife Department, Ministry of Lands and Forestry.

Oates, J.F. (1998) *Myth and Reality in the Rainforest*. Chicago: University of Chicago Press.

Oates, J.F., Whitesides, G.H., Davies, A.G., Waterman, P.G., Green, S.M., Dasilva, G.L. and Mole, S. (1990) Determinants of variation in tropical forest primate biomass: new evidence from West Africa. *Ecology* **71**: 328–343.

Oates, J.F., Abedi-Lartey, M., McGraw, W.S., Struhsaker, T.T. and Whitesides, G.H. (2000) Extinction of a West African red colobus monkey. *Conservation Biology* 1**4**: 1526–1532.

ODI (2001) *Community Forestry in Cameroon*. Rural Development Forestry Network Mailing 25 (16 papers). London: ODI.

Oksanen, T. and Mersmann, C. (2003) Forestry in poverty reduction strategies: an assessment of PRSP processes in sub-Saharan Africa. In *Forests in Poverty Reduction Strategies: Capturing the Potential*, Oksanen, T., Pajari, B. and Tuomasjukka, T. (eds.). *EFI Proceedings* No. 47, 203 pp.

Peres, C.A. (1990) Effects of hunting on western Amazonian Primate communities. *Biological Conservation* 54: 47–59.

Peters, C.M. (1996) *The Ecology and Management of Non Timber Forest Resources*. World Bank Technical Paper 322. Washington, DC: World Bank.

Peters, R.H. (1983) *The Ecological Implications of Body Size*. Cambridge: Cambridge University Press.

Pigeonniere, A. (2001) *Atlas de Guinea Equatorial*. Madrid: Les Editions J.A.

Plumptre, A.J. and Grieser Johns, A. (2001) Changes in primate communities following logging disturbance. In *The Cutting Edge: Conserving Wildlife in Logged Tropical Forests*, Fimbel, R.A., Grajal, A. and Robinson, J.G. (eds.). New York: Columbia University Press, pp. 71–92.

Poulsen, J.R., Clark, C.J. and Malonga, R. (2005a) *Recensement et Distribution des Grands Mammifères et Activités Humaines dans l'Unité Forestière d'Aménagement de Kabo*. Report to the WCS, CIB, FFEMand AFD. Kabo, République du Congo-Brazzaville.

Poulsen, J.R., Clark, C.J. and Malonga, R. (2005b) *Recensement et Distribution des Grands Mammifères et Activités Humaines dans l'Unité Forestière d'Aménagement de Loundoungou*. Report to the WCS, CIB, FFEMand AFD. Kabo, République du Congo-Brazzaville.

Poulsen, J.R., Clark, C.J. and Malonga, R. (2005c) *Recensement et Distribution des Grands Mammifères et Activités Humaines dans l'Unité Forestière d'aménagement de Pokola*. Report to the WCS, CIB, FFEMand AFD. Kabo, République du Congo-Brazzaville.

Rao, M. and McGowan J.K. (2002) Wild meat use, food security, livelihoods and conservation. *Conservation Biology* **16**: 580–583.

Reid, J., Morra, W., Posa Bohome C. and Fernandez Sobrado, D. (2005) *The economics of the primate trade in Bioko, Equatorial Guinea*. Santa Cruz and Washington, DC: Conservation Strategy Fund and Conservation International.

Richards, P. (1985) *Indigenous Agricultural Revolution: Ecology and Food Production in West Africa*. London: Unwin Hyman.

Roberton. S, Huyen Van Thuong, Nguyen Ngoc Nguyen, Ho Loi, Le Hoang Son, Nguyen Quyen, Vu Ngoc Anh, Le Van Di, Barney Long, Vu Ngoc Thanh and Hoang Xuan Thuy (2004) *The Illegal Wildlife Trade in Quang Nam Province: Covert Investigations by Specially Trained Forest Rangers. Wildlife Law Enforcement Strengthening*. Report No. 5. Hanoi, Vietnam: The Quang Nam Forest Protection Department and WWF Indochina.

Roberts, A.D. (1973) *A History of the Bemba: Political Growth and Change in North-Eastern Zambia before 1900*. Madison, WI: University of Wisconsin Press.

Roberts, C.M. (1997) Ecological advice for the global fisheries crisis. *Trends in Ecology and Evolution* **12**: 35–38.

Robertson, C.C. (1984) *Sharing the Same Bowl: a Socioeconomic History of Women and Class in Accra, Ghana*. Bloomington, IN: Indiana University Press.

Robinson, J.G. and Bennett, E.L. (2000) Carrying capacity limits to sustainable hunting in tropical forests. In *Hunting for Sustainability in Tropical Forests*, Robinson, J.G. and Bennett, E.L. (eds.). New York: Columbia University Press, pp. 13–30.

Robinson J.G. and Bennett E.L. (2002) Will alleviating poverty solve the bushmeat crisis? *Oryx* **36**: 332.

Robinson, J.G. and Bennett, E.L. (2004) Having your wildlife and eating it too: an analysis of hunting sustainability across tropical ecosystems. *Animal Conservation* **7**: 397–408.

Robinson, J.G. and Redford, K.H. (1994) Measuring the sustainability of hunting in tropical forests. *Oryx* **28**: 249–256.

Robinson, J.G., Redford, K.H. and Bennett, E.L. (1999) Wildlife harvest in logged tropical forests. *Science* **284**: 595–596.

Roe, D. (2001) Community-based wildlife management: improved livelihoods and wildlife conservation. Biology Briefs No.1. London: International Institute for Environment and Development.

Rose, A.L. (2001) Social changes and social values in mitigating bushmeat commerce. In *Hunting and Bushmeat Utilisation in the African Rainforest: Perspectives Towards a Blueprint for Conservation Action*, Bakarr, M.I., Fonseca, G.A.B.D., Mittermeier, R.A., Rylands, A.B. and Painemilla, K.W. (eds.). Washington, DC: Conservation International, pp. 59–74.

Rose, C.M. (2000) Common property, regulatory property, and environmental protection: comparing community-based management to tradable environmental allowances. In *The Drama of the Commons*:, Ostrom, E., Dietz, T., Dolsak, N., Stern, P.C., Stonich, S. and Weber, E.U. (eds.). Washington, DC: National Academy Press, pp. 233–257.

Ros-Tonen, M.A.F. (1999) Introduction: NTFP research in the Tropenbos programme. In *Seminar Proceedings. NTFP Research in the Tropenbos Programme: Results and Perspectives*, Ros-Tonen, M.A.F. (ed.) Wageningen, the Netherlands: Tropenbos Foundation, pp. 15–32.

Rowcliffe, J.M., Cowlishaw, G. and Long, J. (2003) A model of human hunting impacts in multi-prey communities. *Journal of Applied Ecology* **40**: 872–889.

Rowcliffe, J.M., de Merode, E. and Cowlishaw, G. (2004) Do wildlife laws work? Species protection and the application of a prey-choice model to poaching decisions. *Proceedings of the Royal Society of London Series B Biological Sciences* **271**: 2631–2636.

Rowcliffe, J.M., Milner-Gulland, E.J. and Cowlishaw, G. (2005) Do bushmeat consumers have other fish to fry? *Trends in Ecology and Evolution* **20**: 274–276.

Rowlands, M. and Warnier, J.P. (1988) Sorcery, power and the modern state in Cameroon. *Man* **23**: 118–132.

Ruiz Pérez, M., Belcher, B., Achdiawan, R., Alexiades, M., Aubertin, C., Caballero, J., Campbell, B., Clement, C., Cunningham, T., Fantini, A., de Foresta, H., García Fernández, C., Gautam, K.H., Hersch Martínez, P., de Jong, W., Kusters, K., Kutty, M.G., López, C., Fu, M., Martínez Alfaro, M.A., Nair, T.R., Ndoye, O., Ocampo, R., Rai, N., Ricker, M., Schreckenberg, K., Shackleton, S., Shanley, P., Sunderland, T. and Youn, Y. (2004) Markets drive the specialization strategies of forest peoples. *Ecology and Society* **9(2)**: 4.

Runge, C.F. (1981) Common property externalities: isolation, assurance and resource depletion in a traditional grazing context. *American Journal of Agricultural Economics* **63**: 595–606.

Runge, C.F. (1984) Institutions and the free rider: the assurance problem in collective action. *Journal of Politics* **46**: 154–181.

Sabater-Pi, J. and Groves, C. (1972) The importance of higher primates in the diet of the Fang of Rio Muni. *Man* **7**: 239–243.

Schenck, M., Nsame Effa, E., Starkey, M., Wilkie, D., Abernethy, K., Telfer, P., Godoy, R. and Treves, A. (2006) Why people eat bushmeat: results from two-choice, taste tests in Gabon, Central Africa. *Human Ecology* **34**: 433–445.

Schreckenberg, K., Marshall, E., Newton, A., te Velde, D.W., Rushton, J., and Edouard, F. (2006) *Commercialisation of Non-timber Forest Products: What Determines Success?* Forest Policy Briefing 10. London: ODI.

Schulte-Herbrüggen, B. and Davies, G. (2006) *Wildlife Conservation and Tropical Timber Certification.* Conservation Report No. 6. London: Zoological Society of London.

Schwartzman, S. (1992) Social movements and natural resource conservation in the Brazilian Amazon. In *The Rainforest Harvest: Sustainable Strategies for Saving the Tropical Forests?*, Counsell, S. and Rice, T. (eds.). London: Friends of the Earth, pp. 207–212.

Scoones, I., Melnyk, M. and Pretty, J. (1992) *The Hidden Harvest: Wild Foods and Agricultural Systems: A Literature Review and Annotated Bibliography.* London: International Institute for Environment and Development.

Shanley, P., Pierce, A.R., Laird, S. and Guillen, A. (2002) *Tapping the Green Market: Certification and Management of Non-timber Forest Products.* People and Plants Conservation Series. London: Earthscan Publications.

Shepherd, C. (1999) Export of live freshwater turtles and tortoises from North Sumatra and Riau, Indonesia: a case study. In *Asian Turtle Trade: Proceedings of a Workshop on Conservation and Trade of Freshwater Turtles and Tortoises in Asia*, van Dijk, P., Stuart, P. and Rhodin, A. (eds.). Lunenberg, MA: Chelonian Research Foundation, pp. 112–119.

Solly, H. (2003) *Vous êtes Grands, Nous Sommes Petits. The Implications of Bulu History, Culture and Economy for an Integrated Conservation and Development Project (ICDP) in the Dja Reserve, Cameroon.* PhD Thesis, Université Libre de Bruxelles, Belgium.

Solly, H (2004) *Bushmeat Hunters and Secondary Traders: Making the Distinction for Livelihood Improvement.* Wildlife Policy Briefing No. 8. London: ODI.

Srikosamatara, S., Siripholdej, B. and Suteethorn, V. (1992) Wildlife trade in Lao P.D.R. and between Lao P.D.R. and Thailand. *Natural History Bulletin of the Siam Society* **40**: 1–47.

Starkey, M. (2004) *Commerce and Subsistence: the Hunting, Sale and Consumption of Bushmeat in Gabon.* PhD Thesis, University of Cambridge.

Steel, E.A. (1994) *Study of the Value and Volume of Bushmeat Commerce in Gabon.* Libreville, Gabon: WWF Programme pour le Gabon.

Stephens, D.W. and Krebs, J.R. (1986) *Foraging Theory.* Princeton: Princeton University Press.

Suich, H. (2003) *Summary of Partial Results from the Socio-economic Household Survey Regarding Community-based Natural Resource Management and Livelihoods in Caprivi and Kunene.* WILD Project Working Paper No. 12. Windhoek, Namibia: WILD.

Taylor, K.D. (1961) *An Investigation of Damage to West African Cocoa by Vertebrate Pests.* Unpublished report, W.A. Commonwealth Governments, MAFF, UK.

Teale, A.J. (2003) *Novel Immunisation Strategies against Protozoan Parasites.* Nairobi, Kenya: International Livestock Research Institute.

Te Velde, D., Rushton, J., Schreckenberg, K., Marshall, E., Edouard, F., Newton, A. and Arancibia, E. (2006) Entrepreneurship in value chains of non-timber forest products. *Forest Policy and Economics* **8**: 725–741.

Turrell, G. (1998) Socioeconomic differences in food preference and their influence on healthy food purchasing choices. *Journal of Human Nutrition and Dietetics* **11**: 135–149.

Tutu, K.A., Ntiamoa-Baidu, Y. and Asuming-Brempong, S. (1996) The economics of living with wildlife in Ghana. In *The Economics of Wildlife: Case Studies from Ghana, Kenya, Namibia and Zimbabwe*, Bojo, J. (ed.). Washington, DC: World Bank, pp. 11–37.

Unwin, A.H. (1909) *Report on the Forests and Forestry Problems in Sierra Leone*. London: Waterlow.

Vaughan, C., Katjiua, J., Mulonga, S. and Branston, N. (2003a) *Living with Wildlife: Proceedings of a Workshop to Evaluate Wildlife Utilisation and Human conflict with Community Game Guards*. Kunene. WILD Project working paper 16. Windhoek, Namibia: WILD.

Vaughan, C., Katjiua, J. and Mulonga, S. (2003b) *Cash from Conservation: a Short Survey and Review of the Torra Conservancy Cash Payouts to Individually Registered Members*. WILD Project working paper 15. Windhoek, Namibia: WILD.

Vaughan, C., Katjiua, J., Bundra, K. and Branston, N. (2003c) *Proceedings of a Workshop Held to Discuss WILD Project Key Findings and Build Stakeholder Consensus*. Ombinda Lodge. Kunene. WILD Project working paper 18. Windhoek, Namibia: WILD.

Vaughan, C., Katjiua, J., Mulonga, S. and Murphy, C. (2004) Wildlife use and livelihoods. *In Livelihoods and CBNRM in Namibia: The Findings of the WILD Project,* Long, A. (ed.). Prepared for the Directorates of Environmental Affairs and Parks and Wildlife Management, the Ministry of Environment and Tourism of the Government of the Republic of Namibia, Windhoek, pp. 81–104.

Viljoen, P.J. (1982) *Western Kaokoland, Damaraland and the Skeleton Coast Park Aerial Game Census*. Report prepared for the Namibia Wildlife Trust. Windhoek, Namibia.

von Bubnoff, A. (2005) Africa urged to create more fish farms. *Nature* **436**: 1077.

Watson, I. and Brashares, J.S. (2004) *The Bushmeat Trade and Fishing Licence Agreements in West Africa*. Wildlife Policy Briefing No 4. London: ODI.

WCS (2006) Projet Gibier. http://www.wcsgabon.org/Gibier/BushMeat.html.

WCS and Sarawak Forest Department (1996) *A Master Plan for Wildlife in Sarawak*. Kuching, Sarawak. Wildlife Conservation Society and Sarawak Forest Department.

WCS and TRAFFIC (2007) *Hunting and Wildlife Trade in Asia: Proceedings of a Strategic Planning Meeting of the Wildlife Conservation Society and TRAFFIC, Bali, Indonesia, August 2004*. Kuala Lumpur, Malaysia: WCS and TRAFFIC (in press).

Whitesides, G.H., Oates, J.F., Green, S.M. and Kluberdanz, R.P. (1988) Estimating primate densities from transects in West African rain forest: a comparison of techniques. *Journal of Animal Ecology* **57**: 345–367.

WHO (2004) *Geographical distribution of African Trypanosomiasis, Communicable Disease Surveillance and Response (CSR)*. Geneva: World Health Organization; available at http://www.who.int/emc/diseases/tryp/trypanogeo.html (accessed 12 June 2004).

Wilkie, D. (1989) Impact of roadside agriculture on subsistence hunting in the Ituri forest of Northeastern Zaire. *American Journal of Physical Anthropology* **78**: 485–494.

Wilkie, D.S. (2001) Bushmeat trade in the Congo Basin. In *Great Apes and Humans*, Beck, B.B., Stoinski, T.S., Hutchins, M., Maple, T.L., Norton, B., Rowan, A., Stevens, E.F. and Arluke, A. (eds.) Washingto, DC: Smithsonian Institution Press.

Wilkie, D.S. and Carpenter, J.F. (1999) Bushmeat hunting in the Congo Basin: an assessment of impacts and options for mitigation. *Biodiversity and Conservation* **8**: 927–955.

Wilkie, D.S. and Godoy, R.A. (2001) Income and price elasticities of bushmeat demand in lowland Amerindian societies. *Conservation Biology* **15**: 1–9.

Wilkie, D.S., Starkey, M., Abernethy, K., Effa Nsame, E., Telfer, P. and Godoy, R. (2005) Role of prices and wealth in consumer demand for bushmeat in Gabon, Central Africa. *Conservation Biology* **19**: 268–274.

Wilks, A. and Lefrançois, F. (2002) *Blinding with Science or Encouraging Debate? How World Bank Analysis Determines PRSP Policies*. World Vision and Bretton Woods Project (available: at www.brettonwoodsproject.org/topic/adjustment/blinding/blindful.pdf).

Wollenberg, E. and Ingles, A. (1998) *Incomes from the Forest. Methods for the Development and Conservation of Forest Products for Local Communities*. Bogor, Indonesia: Center for International Forestry Research.

World Bank (2001) *2001 World Bank Atlas*. Washington, DC: World Bank.

World Bank (2002) *World Development Report 2000/1*. Washington, DC: World Bank.

World Bank (2004) Board presentations of PRSP documents (available at www.worldbank.org/poverty/ strategies).

World Bank and IMF (2002) *Review of the Poverty Reduction Strategy Paper (PRSP) Approach: Main Findings*. Prepared by the Staffs of the World Bank and IMF, Washington, DC.

Wunder, S. (2001) Poverty alleviation and tropical forests – what scope for synergies? *World Development* **29:** 1817–1833.

Index

Note: page numbers in *italics* refer to figures and boxes; those in bold refer to tables. Suffix 'n' refers to notes.